HERITAGE AND CONTINUITY IN EASTERN EUROPE

TRANSYLVANIA IN THE HISTORY OF THE ROMANIANS

CORNELIA BODEA and VIRGIL CÂNDEA

EAST EUROPEAN MONOGRAPHS, BOULDER
DISTRIBUTED BY COLUMBIA UNIVERSITY PRESS
NEW YORK

1982

EAST EUROPEAN MONOGRAPHS CXVII

Copyright © 1982 by Cornelia Bodea and Virgil Cândea
Library of Congress Catalog Number 82—70724
ISBN 0—88033—630—3

Printed in Romania

TABLE OF CONTENTS

Foreword   V

I. Transylvania and the Formation of the Romanian People   1

II. Transylvania at the Time of the Magyars' Arrival in Europe   6

III. Transylvania: A Separate Political Entity   9

IV. The Character of Magyar Rule in Transylvania in the Fourteenth, Fifteenth, and Sixteenth Centuries   15

V. Transylvania after the Dismemberment of the Hungarian Kingdom (1526—1541)   19

VI. The Union of Transylvania with Wallachia and Moldavia in 1600   22

VII. The Consequences of the Establishment of Habsburg Domination over Transylvania   25

VIII. The Demography of Transylvania in the Middle Ages and in Modern Times   28

IX. The Resistance of the Transylvanian Romanians to the Policies of the Magyar Nobility and of the Habsburg Regime in the Sixteenth, Seventeenth, and Eighteenth Centuries   34

X. Horea's Uprising of 1784 and its Historic Significance   38

XI. From *Supplex Libellus Valachorum* (1791) to the Revolution of 1848   42

XII. The Austro-Hungarian Compromise of 1867 and its Impact on Transylvania   52

XIII. The Policy of Magyarization and National Oppression in the Dual Monarchy   55

XIV. The Union of Transylvania with Romania   63

XV.	The Consequences of the Union of Transylvania with Romania	71
XVI.	Magyar Revisionism between the Two World Wars	74
XVII.	The Vienna Diktat and its Significance	77
XVIII.	The Liberation of Northwestern Transylvania	82
XIX.	The Status of Nationalities in Romania	86
XX.	Cultural Policy and the Coinhabiting Nationalities	91
	Appendices	99
	Index	169
	Index of quoted authors and works	176
	List of illustrations	179

FOREWORD

Progress in scientific documentation and development of mass media facilitate, nowadays, the securing of substantial data for anyone desirous of knowing more about lands and peoples, about their geography, history, and civilization. These advances are significant for the establishment of global peace and understanding since knowledge generates respect and friendship and cooperation among peoples.

This summary presentation was prepared for that purpose. It is intended to satisfy the interests and concerns of those who desire to become acquainted with Romania and who wish to know more about the past and present of one of the most beautiful regions of that country, Transylvania, about the life, work, and cultural achievement of its inhabitants, Romanians, Magyars, Germans, and others.

It also was intended to meet the needs and expectations of colleagues who have encouraged us to prepare a study concerning Transylvania and its historic significance in the history of the Romanians. This is particularly true of students of Romanian history in the English-speaking world who are in need of a concise volume for instructional purposes.

The authors wish to express their thanks to those individuals and Institutions which have facilitated the publication of this book, and especially to Professor Stephen Fischer-Galati for his editorial suggestions and corollary activities.

★

Romania is a national state with 22,300,000 inhabitants, according to the 1980 data. Of these, 88.14% are Romanians, 7.9% Magyars, 1.6% Germans and 2.36% other nationalities. Coinhabiting nationalities settled among the Romanians starting with the 11th—12th centuries.

The various areas on which the civilization and culture of the Romanian people were formed, all parts of old Dacia, were given specific names during the Middle Ages such as *Țara Românească* or *Muntenia* (Wallachia) for the region stretching south of the Carpathians to the Danube; *Moldova* (Moldavia) for the territory east of the Carpathians to the Dniester; *Oltenia* (Little Wallachia) for the area between the Olt river, the Mehedinți mountains and the Danube; *Transilvania* for the lands encompassed by the Eastern and Southern Carpathians and the Apuseni mountains; *Banat* for the region between the Mureș plain and the Timiș plain: *Crișana* and *Maramureș*, for the northwestern parts; *Dobrogea* (Dobrudja), for the territory between the Danube and the Black Sea. These names, all of great historic significance, are still used today.

Transylvania is the medieval name of the region with which we are concerned in this volume. It is its oldest name as mentioned in documents dated from as early as 1075, 1111, 1113, and so forth [1]. The name, in Latin, means "The land beyond the forest." It is a large plateau furrowed by the Tîrnave, Mureș and Someș river valleys. It covers an area of 99,837 square kilometers and, as of 1980, it had 7,681,433 inhabitants. Transylvania has a good climate and important natural resources such as iron, nonferrous minerals, salt, coal, and natural gas. Its soil is fertile, favorable for the cultivation of cereals, fruit, grapevine. It has forests and meadows propitious for the development of animal husbandry. Transylvania possesses a modern industry and a mechanized agriculture.

The name of Transylvania is currently used not only in relation to the historic principality "beyond the forest" but also to the regions of northern and western Romania, to wit Maramureș, Crișana, and the Banat.

---

[1] *Documente privind istoria României* (Documents on the History of Romania), Series C *(Transylvania)*. București, 1951, vol. I *(1075—1250)*, documents 1—5, pp. 1—3.

I

TRANSYLVANIA
AND THE FORMATION OF THE ROMANIAN PEOPLE

Transylvania is part of the area in which the Romanian people was formed through the union of the native Getae-Dacians and the Romans established in Dacia, after its conquest, in A.D. 106. One of the main political centers of the Getae-Dacians was located in the region of the Orăştie mountains, in southwestern Transylvania. The Getae-Dacians are mentioned in the writings of ancient historians of as early as the sixth century B.C. Herodotus, in the fifth century B.C., characterized them as "the bravest and most righteous of all Thracians." The Getae-Dacians inhabited a vast territory stretching from Haemus — the Balkan mountains of today — to the Northern Carpathians and from the river Tisa to the western shores of the Black Sea. Their spiritual life revealed an advanced civilization and unitary culture.[1]

Transylvania played a decisive role in the history of the Getae-Dacians, of the Daco-Romans, and of their descendants, the Romanians. The Getae-Dacians under the leadership of their king Burebista, a contemporary of Julius Caesar and an ally of Pompey the Great, established in the first century B.C. a powerful centralized state through the union of all political units in Dacia. Although dismembered for a while after Burebista's death, the Dacian kingdom was reunited by Decebalus (A.D. 87—106). Its capital was Sarmizegetusa, close to today's locality of Orăştie, in southwestern Transylvania.

[1] Many of its aspects have been the object of Mircea Eliade's research work in, for example: *Zalmoxis. The Vanishing God. Comparative Studies in the Religions and Folklore of Dacia and Eastern Europe.* Translated by Willard R. Trask. Chicago and London, 1972, X + 260 pp.

Political and military interests, such as the subjugation of the Dacians whose military power prevented Roman expansion to the northeast, as well as economic interests, such as the occupation of a wealthy land, led to a series of military campaigns by the Roman Empire against the Dacians which began as early as the first century A.D. In 106, following two fierce wars, part of Dacia was conquered by Emperor Trajan and transformed into an imperial province the center of which became Ulpia Traiana Sarmizegetusa, located near the old capital whose name was retained. One century earlier, the Dobrudja had been incorporated into the Roman Empire and became a center wherefrom Roman influence could be exerted on the Dacians. The Romans regarded the subjugation of the Dacians as a major political achievement as well as a military and economic one. Thus, in commemoration of the conquest, the Romans erected two monuments: one, Trajan's Column, in Rome and the other, the *Tropaeum Traiani*, in today's Adamclisi in southeastern Romania.

Following to Roman conquest, in the second and third centuries, osmosis occurred between the native Getae-Dacian civilization and Roman civilization, between Getae-Dacians and Romans both in occupied territories and in those of the free Dacians in northern Dacia. That process continued also after the 271—275 period when, under pressure from nomadic peoples' invasions, the Roman armies and administration were withdrawn by Emperor Aurelian south of the Danube. The native inhabitants, strongly Romanized, continued to live on ancestral lands. They retained, to a large extent, their habitual way of life as peasants and shepherds, and underwent ethnic, socio-economic, and cultural changes on Dacian territory completing, in the process, the formation of the Romanian people and of the Romanian language.

The period following the withdrawal of the Roman administration is characterized by a return to village communities as forms of socio-political organization of the native population. However, the old towns and market-towns of Roman Dacia continued to house Daco-Romans. Archeological data are of primary importance in providing evidence for the historic evolution of the population and the development of continuous material and spiritual existence on Dacian territory.

In Transylvania, Daco-Roman remnants of this period have been discovered in burial grounds containing objects characteristic of that civilization through the first half of the sixth century. Such burial grounds of cremation or inhumation are to be found in large numbers in the counties of Cluj, Sibiu, Mureș, Alba and others. Moreover, the character of coins of that period reveals the existence of intensive economic activity by the Daco-Roman population of Transylvania and the Banat inasmuch as the coins in question are Roman silver and bronze coins ranging in time from Aurelian to the beginning of the fifth century which, therefore, were in use even after the withdrawal of the Romans. Many bronze, silver, and gold coins are still being discovered near former Roman settlements in Dacia. After Aurelian's withdrawal Dacia continued to be part of the political, economic, and cultural sphere of the Roman Empire. That empire, always hoping to reconquer the regions which it was forced to abandon, retained several fortified towns on the left bank of the Danube. Those genuine bridgeheads, which the empire did not give up for almost three centuries, also played the role of centers for spreading Roman civilization into the Dacian area. Thus, the Romanian people, from its inception, was incorporated in the civilization of the Roman and later Eastern Roman Empire, a civilization which evolved by taking into account local cultural traditions.

During that period Christianity, which penetrated Dacia during the era of Roman domination, continued to spread in its popular and Latin form. This may be ascertained from archeological finds such as religious objects and buildings, graves and inscriptions as well as by the fact that the fundamental notions of Christianity are expressed in Romanian by words of Latin origin such as *biserică* (basilica), *duminică* (dies Dominica), *Dumnezeu* (Domine-Deus), *cruce* (crucem), *a cumineca* (communicare), *creștin* (christianus), *rugăciune* (rogationem), *a se închina* (inclinare), *păcat* (peccatum), *a răposa* (repausare), *tîmplă* (templum), *Paști* (Paschae). In several parts of Transylvania Christian monuments and objects of the fourth, fifth, and sixth centuries, which belonged to the native Daco-Roman population, have been uncovered. They are found primarily in former urban centers at Napoca (Cluj), Apulum (Alba Iulia), Potaissa (Turda) but also in rural places and around former forts.

The people of Daco-Roman culture developed an ethnic and linguistic identity and a civilization of their own throughout the area they inhabited in old Dacia. Archeological remnants of the sixth and seventh century reveal the forms of life and material culture which were to be found in the civilization of the Romanian people. Written sources of the eighth and ninth centuries in turn attest to the existence of this people.[2] It was organized in village communities, engaged in agriculture and animal husbandry, and possessed remarkable knowledge of iron metallurgy. Ceramics of this period, produced on the territory of old Dacia, reveals uniform characteristics in technique as well as in decoration.

The formation and evolution of the Romanian people could not be arrested by the invasions of the nomadic peoples such as the Visigoths, Vandals, Huns, Gepids, Avars, Slavs, Magyars, Pechenegs, Cumans, and Tatars, which occurred on Dacian territory until the middle of the thirteenth century when earlier Romanian political organizations in Transylvania, Wallachia and Moldavia were established as states. The invasions of the nomadic peoples while slowing down the natural evolution of the native society could not, however, alter the ethnic structure of the Romanian people, its way of life, or the Latin character of its language. The Romanian language, in its vocabulary, has been affected by cohabitation or contacts with other peoples, by borrowings from Slavic, Magyar, German, Turkish, and Greek but has retained its basic character as a Romance language.

Starting in the ninth and tenth centuries Byzantine, Armenian, German, and Russian *(Nestor's Chronicle)* sources together with Hungarian ones refer to Romanians as *Vlahi, Blasi, Blaci, Blachi,*

---

[2] Cf. Mehmet Ali Ekrem, "O mențiune despre românii din secolul al IX-lea în *Oguzname* — cea mai veche cronică turcă" (A Reference to the Romanians of the 9th Century in *Oguzname* — the Oldest Turkish Chronicle), in *Studii și cercetări de istorie veche și arheologie*, 31, 1980, 2, p. 287—294.

Volohi, Balak, Walachen, Oláh in the Carpathian-Danubian-Pontic region clearly distinguishable from other Southeast European peoples. These appelations, derived from *Walk* — the German term for Romanic peoples —, show that the authors of those historic sources were aware of the Roman origins of the Romanians.

II

TRANSYLVANIA
AT THE TIME OF THE MAGYARS' ARRIVAL IN EUROPE

The Magyars, a Finno-Ugric people which originated in Asia in the area between the Altai mountains and northern Iran, under pressure from Pechenegs and Bulgars left the region north of the Black Sea — Etelköz — where they had settled temporarily and, between the years 896 and 900, crossed the Carpathians north of Transylvania under the leadership of Arpad and settled in the Pannonian plain, today's Hungary. In years to follow they attempted to penetrate into Western Europe. Unable to do so they turned toward Transylvania where they met with resistance from local Romanian political units.

'P. Dictus Magister,' the anonymous notary (Anonymus) of King Béla of Hungary records in his chronicle *Gesta Hungarorum* the existence of three "duchies" (voevodeships) of Romanians, one in the region of the Criş rivers (in the west of today's Romania) led by Menumorut, a second in the Banat, ruled by Glad, and a third "beyond the forest" (to the northeast of the Apuseni mountains), between the Meseş pass and the sources of the Someş river, where Gelu *(Gelou dux Blachorum)* was in command.

The inroads from Pannonia eastward occurred in several successive stages between the tenth and the thirteenth centuries. They began through Bihor or the region of the Criş rivers. Anonymus speaks of Menumorut's unwillingness to render the country to "Duke Arpad" and of the reasons for his categoric refusal, i.e. that he had inherited the land from his ancestors and that he was under the protection of the Byzantine emperor. Following repeated confrontations, the natives and the invaders reached an agreement by which the Bihor voevode was able to maintain, in part, his rule. After a while another Magyar leader, Tuhutum, upon learning "from local people of the wealth of the Ultrasylvanian land which was ruled by a Romanian by the name of Gelu" *(dum cepisset audire ab incolis bonitatem terre Ultrasilvane, ubi Gelou quidam Blacus dominium tenebat)*, crossed the Apuseni mountains through the Meseş pass. The Romanian voevode fought Tuhutum at the "Meseş Gates" but was forced to withdraw inland along the rivers Almaş and Căpuş. He was killed in his fortress on the river Someş. Tuhutum succeeded him with the assent of the Romanians who had been under the authority of Gelu. According to the same Anonymus, in the Banat Glad tried to resist with an army consisting of "Cumans, Bulgars, and Blachs." The army, however, was defeated and he surrendered voluntarily.

Anonymus recalls that the Blachs (Romanians) lived also in Pannonia. Specifically, when the Magyars were still north of the Carpathians, in the Land of the Ruthenians, the "dukes" of the Ruthenians encouraged the Magyars to cross the mountains into Pannonia which was inhabited by *Sclavi, Bulgarii et Blachii ac pastores Romanorum*.

The military resistance by the Romanians, led by Menumorut, Gelu, and Glad, against the Magyar invasion reveals not only their existence in significant numbers but also their past characterized by economic and political development. In an original phase, the aforementioned "duchies" retained their individual characteristics. Chronicles of the twelfth, thirteenth, and fourteenth centuries (Simon of Kéza, *Chronicon Monacense, Chronicon Posoniense, Legenda Sancti Gerhardi*) attest to the existence of Romanian political entities in Transylvania and the Banat. At the beginning of the eleventh century, for instance, rulers such as Ahtum or Ohtum, in the Banat, and Gyla in Gelu's former voevodeship, engaged in a policy of total independence from Hungary which brought on the military confrontation with King Stephen I (the Saint) and their surrender. In Transylvania, however, there were also other *ţări (terrae), cnezate* (princedoms), voevodeships or duchies led by Romanian princes or voevodes — native Romanian political entities such as Ţara Făgăraşului, Ţara Amlaşului, Ţara Haţegului, Ţara Maramureşului — which were not mentioned by Anonymus because they were not adjacent to Magyar tribes in the period of early incursions. They continued to exist, safeguarding their old political, juridical, and military traditions, for a long time.

Magyar domination over Transylvania was introduced gradually and with difficulty. The administrative, juridical and military organization, centering on the *comitatus*, could be initiated only in the twelfth century and be completed only in the fourteenth, still only as far as the valleys of the Mureş and of the Tîrnave rivers. Beyond the valleys, the Szeklers and the Saxons, settled or colonized in the twelfth century, were to organise themselves in their own seats *(sedes)*.

III

TRANSYLVANIA: A SEPARATE POLITICAL ENTITY

The voevodeship, the original form of political organization in Transylvania, was also characteristic of the Romanian people of Moldavia and Wallachia.[1] Just as in these Romanian lands, the Transylvanian voevode assumed control over the military, administrative, and judicial functions of the state. He enjoyed complete power, again just as on the other side of the Carpathians, but he did not call himself "sole ruler." He was dependent on the Hungarian crown but that dependency did not alter the basic autonomy of the voevodal structure. Except for Croatia, no similar territorial and political organization was to be found in the old Hungarian kingdom. The *comitatus*, taken from the Carolingians by the Magyars and used in part also in Transylvania did not replace the voevodeship.

As a matter of fact, on their arrival in Pannonia the number of Magyars was too small to allow them to populate or numerically dominate the territories which they occupied. The historian Eckhart Ferenc shows that "the conquerors did not fully populate even the territory which was to become Hungary."[2] They settled in the valleys of the Tisa and of the Danube and in the regions beyond the Danube and the Drava in tribal groups and not in compact areas.[3] All the less were they thus able to impose a new organization either by force of numbers or by ability to do so. Their rule was mostly

---

[1] *Voevodeship* is Slav as a linguistic term but its institutional content is different. Although the term *voevode* meant *minor high official* in the Slav world, the Romanians used it for *supreme chief*, and this specific Romanian form was not found anywhere else.

[2] Eckhart Ferenc, *Storia della nazione ungherese*. Translated by Rodolfo Mosca. Milano 1929, p. 37.

[3] *Ibid*.

nominal. It was only normal therefore for them to retain the traditional Romanian forms of organization which they found; nor is it surprising that the first tendencies toward Christianity displayed by the Magyars were related to the Byzantine church. Gyla accepted the eastern rites when he converted to Christianity, the same rites as those of the Romanian population. Similarly Ahtum, Glad's successor, was christened at Vidin in the Eastern rite.

Even after acceptance of Catholicism by the Magyars, the latter were unable to impose it on the Transylvanian Romanians. The Hungarian kings were unable to fulfil the apostolic obligations they had assumed, which entailed the spreading of Catholicism among the Orthodox, nor were they able to effectively administer or assimilate the territories which they dominated. This is the reason why they were forced to colonize such territories with Szeklers and Saxons and to invite orders of knight-monks both for defending the eastern frontier and for advancing Catholic propaganda.

The Szeklers,[4] who were previously located in western Transylvania, in Bihor, moved during the twelfth century into eastern and, later, southeastern Transylvania, where we still find them today. Their origin as well as the circumstances surrounding their settlement in the Carpathian region are still subject to discussion. They are supposed to be followers of the Avars, descendants of certain Turkic peoples, or of certain tribes which removed themselves from the empire of the Khazars during the ninth century, or related to the Magyars. The Szeklers retained throughout the Middle Ages as well as in modern times their way of life and organization, the *Szekler seats*. They fought repeatedly together with the Romanians against Magyars, Tatars, and Ottoman Turks. Thus, the Moldavian voevode Stephen the Great could count in his armed forces that fought the Turks at Vaslui, in 1475, some 5,000 Szeklers, while Petru Rareș, also a Moldavian voevode, was assisted by the Szeklers, in his Transylvanian campaign of 1541. In 1599—1600 the Szeklers

---

[4] Demény Lajos, "Scurtă privire istorică pînă la revoluția din 1848" (Brief Historic Overview up to the 1848—1849 Revolution) in *Naționalitatea maghiară în România* (The Magyar Nationality in Romania), ed. by Koppándi Sándor, Bucharest, 1981, pp. 30—59.

fought under the command of Michael the Brave, the voevode of Wallachia, against the Magyar Cardinal-Prince Andrew Báthory who had abused their privileges.

Germans were brought in group by group from different regions.[5] Some came from Flanders — the so-called *Flandrenses* —, others came from the valley of the Moselle and from Luxemburg *(Teutonici)*, while some more came from Saxony, Westphalia, Hesse, Thuringia, and Bavaria. The name by which all these groups came to be known is that of Saxons *(Saxones)*. The Germans settled primarily in the Sibiu, Bistriţa, and Rodna regions and, somewhat later, in Ţara Bîrsei, in the southern and northeastern parts of Transylvania. As they enjoyed special privileges granted by the Arpad dynasty they were able to organize themselves on the basis of districts or seats *(sedes)* and to engage in extensive socio-economic activities which, together with the activities of the Romanians and Magyars, contributed to the general flourishing of the region.

From a geographic and economic point of view Transylvania, in the past, was closer to the Romanian principalities than to Hungary. The Apuseni mountains and, particularly the once very large swampy Tisa plain helped, to a considerable extent, the maintenance of Transylvania's individuality and its southeastern economic orientation. Consequently, Transylvania's trade relations were far more developed with Moldavia, Wallachia, and the Ottoman Empire than with Hungary and the West. It was possible for caravans to cross the Tisa only at Szolnok and Szeged but they could reach the Romanian principalities more readily since the Carpathians could be crossed through several mountain passes. Furthermore, the distance from the center of Transylvania to the Black Sea is relatively short — approximately 400 kilometers — whereas the distance to the Adriatic, the maritime outlet of medieval Hungary, is over 800 kilometers.[6] It should also be noted that whereas Buda was indeed

---

[5] Cf. Thomas Nägler, *Die Ansiedlung der Siebenbürger Sachsen*. Bucharest, 1979, 250 p.

[6] L. Someşan, *Die Theissebene, eine natürliche Grenze zwischen Rumänen und Magyaren*. Sibiu, 1939, pp. 92—101; I. Moga, "L'Orientation économique de la Transylvanie," *Revue de la Transylvanie*, VI, 1, 1940 p. 74.

the principal crossroads of the commercial network of medieval Hungary, Transylvania never gravitated toward that center remaining instead a distinct physico-geographic entity. The Magyar geographer Cholnoky Jenö, the exponent of that point of view, further believes that "This special situation prevailed throughout history. Both as Dacia and as Principality of Transylvania this territory always had its own history."[7]

It should also be observed that in the fourteenth and fifteenth centuries certain parts of Transylvania found themselves, temporarily, under the official, de facto, administration of Wallachia (such as Amlaş, Făgăraş, Haţeg) or of Moldavia (Ciceu and Cetatea de Baltă) which, in turn, led to even closer relations between the Romanians on either side of the Carpathians. And this was true even after these territorial entities ceased to be part of the feudal possessions of the trans-Carpathian Romanian rulers. It should be noted, however, that in the sixteenth century the voevode of Moldavia, Petru Rareş, had actual possession of the Unguraşul fortress and the towns of Bistriţa and Rodna with all dependent villages.

The minor significance of commercial relations between Transylvania and medieval Hungary is also revealed by the fact that Transylvania had a different currency, a different system of weights and measures, and different custom duties than Hungary. In Hungary, the primary currency was the Hungarian mark, whereas in Transylvania it was the Sibiu mark.[8] Even during the period of Austrian domination and also under the dualist Austro-Hungarian regime, as late as 1874, the local Transylvanian mark circulated alongside the official Viennese mark.[9]

During the period of Ottoman domination, when central and southern Hungary became pashaliks and Transylvania was recognized as an autonomous principality in 1541, Transylvania's trade focussed more and more on the south and east. The commer-

[7] Cholnoky Jenö, "Budapest földrajzi helyzete" (The Geographic Location of Budapest), *Földrajzi Közlemények*, XLIII, V, 1915, p. 206.

[8] Homan Bálint, *Magyar pénztörtenei* (Magyar Monetary History). Budapest, 1916, pp. 94—100.

[9] *Ibid*; Moga, *L'Orientation économique*, pp. 79—80.

cial outlets and the acquisition of supplies revolved in Moldavia, Wallachia, and the Ottoman Empire.

Following the establishment of Austrian rule after the Peace of Karlowitz (1699), Hungary and Transylvania — the latter now a Great Principality — were incorporated into the realm of autarchic mercantilism practiced by Vienna. While Hungary retained its position as outlet for Bohemia's and Austria's industry, Vienna's industrial policies were such as to diminish the significance of Transylvania's industry. However, the customs exactions imposed on Transylvania's trade with the Romanian principalities and with southeastern Europe were, eventually, doomed to total failure. The interdependence between Transylvania and Moldavia and Wallachia proved to be more powerful than all customs barriers. In 1779, the Transylvanian Court Chancellery in Vienna was forced to acknowledge the "close relationship between the Great Principality of Transylvania and the neighboring provinces of Moldavia and Wallachia" and the disastrous effects of the restrictions.[10] Thus, in following years Transylvania's commerce regained its former status. The balance of Transylvanian foreign trade for 1837—1838 speaks for itself:[11]

		Romanian Principalities and Turkish Provinces	Hungary and Austrian Provinces
1837	Export	2,947,169 Florins	64,569 Florins
1837	Import	3,186,835 Florins	285,199 Florins
1838	Export	2,995,091 Florins	89,144 Florins
1838	Import	4,157,055 Florins	173,408 Florins

Even at the end of the nineteenth century many inhabitants of Transylvania were unfamiliar with Hungary's geography but were cognizant of Wallachia's. A Magyar writer stated that the Szeklers

---

[10] I. Moga, "Politica economică austriacă și comerțul Transilvaniei în veacul XVIII" (Austrian Economic Policy and Transylvania's Trade in the Eighteenth Century), *Anuarul Institutului de istorie națională*, Cluj, VII (1935—1938), pp. 94—96, 139—144; Moga, *L'Orientation économique*, pp. 94—98.

[11] Moga, *Politica economică*, p. 96.

believed Hungary to be as mountainous as Ciuc and Trei Scaune; they only had hearsay knowledge of Szeged but even their children knew of Bucharest, Brăila, or Sinaia.[12]

It was therefore possible to state in 1866 in Magyar historiography that "today nobody denies any longer[...]the separate character of the history of Transylvania,"[13] in many ways different from that of the Hungarian kingdom. In this respect the Szekler chronicler of the eighteenth century, Cserei Mihály, bitterly observed that "the threat for Transylvania always came from Hungary and from Hungarians"[14]:

"At least from now on you should learn Transylvania, my dear country,
That you are living peacefully with your own brethren under one roof;
Do no longer befriend as readily those of Hungary
So that you won't suffer, as you now do, from your own dire harm."[15]

---

[12] Szöke Mihály, *Pusztuló véreink. Adatok a székely kérdéshez* (Our Bloodkin Perishes — Data for the Szekler Question). Budapest, 1902, pp. 12 and 18.

[13] Szilágyi Sándor, *Erdélyország Története* (History of Transylvania). Pest, 1866, vol. I, p. VI.

[14] Nagyajtai Cserei Mihály, *Historiája* 1661 — 1711. Second edition. Pest, n.d.; E. Lukinich, "Les idées politiques dirigeantes de la Principauté de Transylvanie de 1541 à 1600", *Bulletin d'information des sciences historiques en Europe Orientale*, V, 1933, p. 9.

[15] Nagyajtai, *Historiaja*, p. 1.

IV

THE CHARACTER OF MAGYAR RULE IN TRANSYLVANIA IN THE FOURTEENTH, FIFTEENTH, AND SIXTEENTH CENTURIES

Until the extinction of the Arpad dynasty the Hungarian sovereigns did not differentiate among their subjects in terms of religion or ethnic origin. The Romanian voevodes in Transylvania retained elective or hereditary prerogatives as recognized by the Hungarian sovereigns. Moreover, during that period the Romanians participated politically, alongside the Magyars, Szeklers, and Saxons in the diets (or congregations) of Transylvania. Such, for instance, was the general assembly of 11 March 1291, held in Alba Iulia, *cum universis Nobilibus, Saxonibus, Siculis et Olachis in partibus Transilvanis*, related to the revision of the statute of their congregation.[1] This was also the case later, in 1355. The status of the Romanians worsened during the second half of the fourteenth century. Beginning with the year 1351, when the reorganization of the institution *Locus credibilis* (pl.: *Loca credibilia*)[2] officially entrusted Catholic chapters with the preparation of civil and notarial record as well as the authentication thereof, the Orthodox faith was gradually taken outside the law. A crucial act was the royal decree of 1366 whereby all aristocratic landed property was subject to authentication in terms of the owners' fidelity to the king. In other words, the decree entailed the possibility of royal refusal to recognize the rights of possession to nobles who were not Catholics or to those proven unfaithful to the king. Thus the clergy and aristocracy of Orthodox faith,

---

[1] Zimmermann-Werner, *Urkundenbuch zur Geschichte der Deutschen in Siebenbürgen*. Hermannstadt, 1892, vol. I, p. 177.

[2] *Locus credibilis*, archives where authentic texts of juridical documents were kept in the Middle Ages. Magyar: *hiteles hely*. Italian : *luogo credibile*.

that is the Romanians, were excluded from the country's political life.

This deterioration and corollary acceleration of anti-Romanian measures adopted in Transylvania were directly related to the establishment of the Romanian states of Wallachia and Moldavia which successfully resisted encroachment attempts by the Hungarian kingdom. The Angevin king Louis the Great was concerned over the possibility of extension of the emancipation movement which was manifest in Transylvania, especially in Maramureş. In 1365 he ordered the confiscation of the lands of the Maramureş voevode Bogdan, who had become the leader of the resistance movement in Moldavia, so "that this wretched act should not serve as example for others who would attempt similar actions" *(ut ne perversitas aliis similia presumpmentibus transeat in exemplum)*.[3] The following decree of 28 June 1366, issued after Bogdan's movement succeeded in Moldavia, contained measures of extreme repression in Transylvania directed primarily against the Romanians *(specialiter olahi)*.[4]

Taking into account the motive that led to the issue of the document on June 28, 1366 [5] — providing strict measures against the revolted "Wlachs" in Transylvania, while other steps were concomitantly taken to favor the Catholic propaganda and hinder action by the Orthodox clergy in the Banat that was considered schismatic — it is hardly surprising that after that date the presence of

---

[3] I. Mihalyi, *Diplome maramureşene din sec. XIV şi XV* (Maramureş Diplomas in the Fourteenth and Fifteenth Centuries). Sighet, 1900, p. 57; cf. *Documenta Historiam Valachorum illustrantia usque ad annum 1400*. Budapest, 1941, pp. 178—181 (abstract).

[4] Cf. Şerban Papacostea, "La fondation de la Valachie et de la Moldavie et les Roumains de Transylvanie: Une nouvelle source," *Revue Roumaine d'Histoire*, XVII, 3, 1978, pp. 389—407; also Maria Holban "Românii din Transilvania şi anarhia feudală din secolul al XIV-lea" (The Romanians in Transylvania and the Feudal Anarchy in the 14th Century) in *Din cronica relaţiilor româno-ungare în secolele XIII—XIV* (From the Chronicle of Romanian-Hungarian Relations in the 13th—14th Centuries). Bucharest, 1981, pp. 245—300.

[5] Eudoxiu de Hurmuzaki, *Documente privitoare la istoria românilor* (Documents on the History of the Romanians), Bucureşti, 1890, vol. I, p. 120; Zimmerman-Werner Müller, *Urkundenbuch zur Geschichte der Deutschen in Siebenburgen*, Hermannstadt, 1897, vol. II, p. 256.

Romanians at the assemblies of Transylvanian "states" was no longer mentioned. "The existence, at the very boundaries of the [Transylvanian] province, of the two independent or quasi-independent Romanian principalities that were already revolving in the orbit of other powers in order to find support against the Magyar hegemony and whose setting up was the work of rebels from Transylvania, inevitably induced the representatives of the Hungarian royal authority to follow a policy characterized by suspicion and abuse."[6]

This policy of discrimination continued. Under the circumstances, certain Romanian nobles became converted to Catholicism and were Magyarized and, in the process, established leading aristocratic families such as the Kendeffy, Josika, Naláczy, Dragfy, Bizere, and others. But there were also other illustrious figures which did not deny their Romanian origin, such as the Hunyadi (fifteenth and sixteenth centuries), Nicolaus Olahus (sixteenth century), Ioannes Caioni (seventeenth century) and others.

In the fifteenth century, one of the Romanians who reached the highest political and military positions in Hungary, John Corvinus of Hunedoara, was also one of the foremost military leaders of his age. Voevode of Transylvania (1441—1446), governor or regent (1446—1452) and, later, supreme commander of Hungary (1453—1456), John of Hunedoara led the crusade against the Ottoman Turks between 1441 and 1456, in an age when Mohammed the Conqueror took Constantinople. It is noteworthy that as voevode of Transylvania or regent of Hungary John of Hunedoara used primarily Romanian troops in fighting the Turks. The accounts of the great battle of Varna of 1444, for instance, designate the army led by John of Hunedoara as the "army of the Walachs" *(acies Walachorum)*.[7] That army consisted not only of Romanians from Transylvania but also of military forces of the allied Romanian voevodes of Moldavia and Wallachia who fought together in the Carpathians, on the Danube, on the shores of the Black Sea, from Varna to Belgrade, and delayed the Ottoman advance into Central Europe by

---

[6] G. I. Brătianu, "Les Assemblées d'états et les Roumains en Transylvanie," *Revue des Études Roumaines*, Paris, XIII—XIV, 1974, p. 25.

[7] *Monumenta medii aevi historica res gestas Poloniae illustrantia*. Vol. XII. Cracoviae, 1891, p. 464.

more than half-a-century. Two years after John's death, his son, Matthias Corvinus, became king of Hungary. His rule of 32 years was marked by a brilliant cultural and artistic Renaissance.

The large majority of Romanians, who remained Orthodox, were stigmatized. Moreover, the general hardening of feudal exactions, so characteristic of the entire era, affected primarily the Romanians, the majority of the peasantry. Several local uprisings of Romanian and Magyar serfs occurred in Transylvania starting in the fourteenth century. They were to culminate in the following century, in 1437, in the Bobîlna uprising, in the region of the Someş (Dej-Cluj region). In the course of the rebellion, the nobility concluded the Union of Căpîlna, an agreement or fraternal union[8] (*fraternam unionem*) with the Szekler and Saxon leaders by which each side agreed to provide reciprocal aid against internal and foreign enemies. Because the contracting sides were three in number, the union became known as *unio trium nationum*. The union bore fruit to the extent that the peasant rebellion was put down, its leaders were executed, and thousands of peasants were mutilated. The masses' living conditions became even harder.

The Bobîlna uprising represents an important chapter in the history of social revolts *per se* and in terms of Romanian-Magyar collaboration. From a national political standpoint, however, the consequences of the *fraternam unionem* of 1437 were grave and far-reaching. This, because it later became the basis of the theory recognizing only three political and ethnic-territorial nations in Transylvania, to wit the Magyars, the Saxons, and the Szeklers, to the exclusion of the Romanians.[9] Thus, for Romanians, as peasants in general and as Romanians in particular, the union of 1437 became a century-long instrument of social and national oppression.

---

[8] See pp. 20—21.

[9] D. Prodan, *Supplex Libellus Valachorum or The Political Struggle of the Romanians in Transylvania During the Eighteenth Century*. Bucharest, 1971, pp. 64 ff.

V

TRANSYLVANIA AFTER THE DISMEMBERMENT OF THE HUNGARIAN KINGDOM (1526—1541)

The events following the Turkish victory at Mohács (1526) which led to the dismemberment of the Hungarian kingdom, and the occupation of the fortress of Buda by the Ottomans, in 1541, contributed to Transylvania becoming an autonomous principality under Ottoman suzerainty. The wars for the Hungarian kingdom fought between Ferdinand of Habsburg and the voevode of Transylvania, John Zápolya, who enjoyed the support of the Porte, ended in 1541 with the decision of Suleiman the Magnificent to transform central and southern Hungary into a *pashalik* with Buda as its center. Thus, northwestern Hungary, with its center at Pressburg (Pozsony) — Bratislava of today — together with Slovakia, Croatia, and Slavonia remained under Habsburg rule. The Transylvanian principality — which for a time included also the Banat and the area known as *Partium* (Crişana and Maramureş) and had its capital at Alba Iulia — retained its autonomy under Ottoman suzerainty.

The statute granting autonomy to Transylvania shows that the Porte regarded the province as a political entity different from Hungary and, in fact, subjected Transylvania to a regime similar to that of the two Romanian states, Moldavia and Wallachia. It is known that in its expansion into southeastern and central Europe the Ottoman Empire did away with Christian states such as the Bulgarian and the Serbian tsardoms, the Byzantine Empire, the Hungarian kingdom, and others. A different policy was enforced with respect to Moldavia and Wallachia which retained their political autonomy, as recognized by treaties,[1] in return for payments of tributes in cash and in kind. Thus, the political and administrative leadership of the Romanian provinces remained in native hands as both Moldavia and Wallachia were ruled by Romanian voevodes. Except for fortresses on the Danube, there were no Turkish garrisons on Romanian soil; nor were there any mosques or any conversions to Islam. These conditions were significant not only in assuring the continuity of native rule and institutions but also for later political, economic, and military activities emanating from or focussing on the Romanian principalities, designed to advance the gaining of independence by the peoples of the Balkan Peninsula.

Transylvania benefited from the same autonomy it enjoyed throughout the period of Ottoman suzerainty which ended in the late seventeenth century. The principality was ruled by a prince, elected by the Diet and confirmed by the Porte. However, on occasion, the prince could be named directly by the sultan. The prince's prerogatives were similar to those of the rulers of Wallachia and Moldavia. The Transylvanian prince ruled with the assistance of the Council *(Consilium)* and of the Diet. The Diet represented only the privileged groups, those which assumed privileged status in 1437. Thus, in the Transylvanian political order, national discrimination worked side by side with class discrimination specific to feudalism in general. Transylvanian legislative enactments invariably granted rights and privileges only to three political nations, "nobles, Saxons, and Szeklers", and to four recognized religions, Catholic, Lutheran, Calvinist, and Unitarian. By the same token, they consistently excluded from participation in public and political life the largest nation, the Romanians, and the religion of the majority of the population, the Orthodox. The three privileged nations, called as they were *nobles, Saxons,* and *Szeklers,* accordingly divided the land into

---

[1] See Aurel Decei, "Tratatul de pace — sulhnâme — încheiat între sultanul Mehmed al II-lea şi Ştefan cel Mare la 1479" (The Peace Treaty—Sulhnâme— Concluded between Sultan Mehmed II and Stephen the Great in 1479), in his volume *Relaţii româno-otomane* (Romanian-Ottoman Relations). Bucureşti, 1978, p. 139 ff; Ion Matei "Quelques problèmes concernant le régime de la domination ottomane dans les pays roumains," in *Revue des études sud-est européennes*, 10, 1972, no 1, p. 65—81; 11, 1973, no 1, p. 81—95.

three parts, to wit: "land of the nobles," "Saxon land" and "Szekler land". In turn, the "nation of the nobles" in the process of national evolution soon called itself the "Magyar nation," and consequently the "land of the nobles" was claimed as "Magyar land." The Romanians, according to the privileged nations, could not have a country *(terra)* or a land, because 1) they were not recognized as a privileged nation, 2) they did not have a recognized religion, and 3) they were merely tolerated temporarily in Transylvania. In other words, they were denied not only autochthony but also the elementary right conceded to colonists, the land to live on.

The set of bills voted between 1540 and 1653, including the codified discriminations against the Romanians in terms of nation and religion, was called *Approbatae Constitutiones Regni Transilvaniae et Partium Hungariae eidem annexarum* (The Approved Constitutions of the Land of Transylvania and of the Parts of Hungary Annexed to It).[2] The discriminatory bills were drafted in harsh, offending terms, evincing marked hostility particularly after the Romanian rule of the Wallachian prince Michael the Brave in Transylvania (1600) (see pp.22−24). Starting in the eighteenth century, the Romanians objected to such abusive practices, by way of petitions and uprisings. In their memoranda and petitions they claimed equal social and political status with the other privileged political nations and insisted above all on abrogation and annulment by law of all the "hateful and insulting," "unjust and shameful" terms that were "unlawfully stamped on the forehead of the Romanian nation."[3]

---

[2] Printed in 1653.

[3] The quoted passages are reproduced from *Supplex Libellus Valachorum Transilvaniae or The Transylvanian Romanians' Supreme Petition*, drafted in 1791. See Appendix B, p. 134.

VI

THE UNION OF TRANSYLVANIA WITH WALLACHIA AND MOLDAVIA IN 1600

Through the unification into one state of the Romanian people, achieved in 1600 by the ruler of Wallachia Michael the Brave, the temporary remaking of Dacia was accomplished. The struggle for the union of Transylvania with Wallachia and Moldavia took place within a general European historic context, within the framework of the anti-Ottoman coalition initiated by the Papacy in 1592—1593 and patronized by the Habsburg Emperor Rudolf II. The Christian League, to ensure the success of its project, relied heavily on the collaboration of countries which had a long tradition of wars with the Ottoman Empire — *Un esercito formato et confederato col Transilvano, Valacco et Moldavo...* [1]

Michael became ruler of Wallachia in September 1593. The offensive-defensive understanding among Wallachia, Transylvania, and Moldavia was a matter of record by the following year. Michael's leadership during the confrontations both north and south of the Danube singled him out as a military genius, as a man of courage and determination, as a man whose fame surpassed that of other leaders of the Christian alliance. For these reasons, exacerbated by the difficulties and vicissitudes of the campaigns themselves, he had to cope with the intrigues, indecision, and pride of rivals.

To unify the forces of the Romanians and to present a united front against the Ottoman Empire, Michael brought together Wallachia with Transylvania (1599) and Moldavia (1600). As ruler of the three Romanian lands he assumed the title "Michael-Voevode, by the Grace of God Ruler of Wallachia, of Transylvania, and of

---

[1] Hurmuzaki-Iorga, *Documente*, vol. XI, p. 389.

all of Moldavia". Although the union lasted for only a short period of time — from May 1600 until August 1601 — it enhanced, through its political consequences, the consciousness of the Romanians of the three principalities regarding their existence as one people with a common language, common religion, common customs, and common ideals.

The national character of the union of 1600 was perceived also by foreign recorders of history of the time. The Magyar-Transylvanian chronicler Szamosközy István stressed the fact that Michael was well received in Transylvania because the Romanian population regarded him as "one of theirs," in other words, a Romanian ruler. The same views were also held by the Saxon chronicler Georg Kraus. That national character is also evident in the measures adopted by Michael the Brave designed to do away with political, social, and religious segregation of the Romanians of Transylvania which included the naming of Romanian nobles to the Prince's council alongside Magyar and Szekler aristocrats, the placing of Romanian commanders in several Transylvanian fortresses, the use of Romanian in acts of the Prince's chancery, measures favoring Romanian priests and serfs, and the establishment of the seat of the Romanian metropolitan of Transylvania at Alba Iulia, and his hierarchical subordination to the Metropolitan Church of Wallachia.

The unity of the Romanians on the former territory of Dacia was recorded also by all European humanists who were concerned with that people between the fourteenth and seventeenth centuries such as Flavio Biondo, Aeneas Sylvius Piccolomini (Pope Pius II), Poggio Bracciolini, Rafaello Maffei (Volterrano), Nicolaus Olahus, Antonio Bonfini (Matthias Corvinus' secretary) as well as by learned Saxons such as Christian Schaeseus, Georg Reicherstorffer, Laurentius Toppeltinus, Johann Tröster, and many, many others.[2]

In 1574, the Frenchman P. Lescalopier wrote: "This entire land [Wallachia], and Moldavia and most of Transylvania, was inhabited by Roman colonies from the times of Emperor Trajan... The inhabitants of these lands call themselves true descendants of

---

[2] See for instance Adolf Armbruster, *La Romanité des Roumains. Histoire d'une idée*. Bucarest, 1977. p. 279.

the Romans and call their language *romaneschte*, which means Roman."[3] Political leaders, diplomats, and military men of sixteenth- and seventeenth-century Europe considered that the three Romanian lands constituted a political, economic, and military entity in their planning of military campaigns originating in Europe and directed against the Ottoman Empire.

It should also be noted that attempts at unification of all or of some Romanian lands under foreign domination were entertained until as late as the nineteenth century. Such attempts were conceived by the Venetian adventurer in the service of the Turks, Aloisio Gritti (1532), by Grantrye de Granchamps, the French ambassador to Constantinople (1566), by Henri de Turenne, the famous general during the reign of Louis XIV (1633); they were also proposed, repeatedly, by Catherine II of Russia in 1771, 1782, 1783, and 1790. The project of unification was again formulated during the Russo-Turkish war of 1806—1812. The Greek diplomat Capodistrias too suggested, in 1828, the formation of a "duchy or kingdom of Dacia" under Russian protection. Evidently, none of these projects were concerned with the formation of the Romanian national state. But it is clear that a Dacian kingdom—as a buffer state between the three great empires—was envisaged also because a common language, Romanian, would have facilitated its governance and administration.

The union of the Romanian lands achieved by the Wallachian ruler in 1600 was, until the total union of 1918, symbolic of the formation of the unitary national Romanian state. For neighboring empires the name and deed of Michael the Brave have always been a source of concern. The Habsburg Empire, the Austro-Hungarian one of later years, the Ottoman Porte, and the Tsarist Empire were opposed to the attainment of the Romanian national state because they viewed such an accomplishment as an impediment to expansion toward Southeast Europe and as a precedent which could result in the dismemberment of their own heterogeneous empires. For the Porte it threatened the very existence of Ottoman rule in the Balkan Peninsula.

[3] „Le voyage de Pierre Lescalopier Parisien, de Venise à Constantinople l'an 1574", *Revue d'histoire diplomatique*, XXXV, 1921, no. 1.

VII

THE CONSEQUENCES OF THE ESTABLISHMENT OF HABSBURG DOMINATION OVER TRANSYLVANIA

Between 1682 and 1699 a new anti-Ottoman war started by Austria and Poland, who were joined in a Holy League in 1684 by the Papacy, Venice and, later, also Russia, led to the removal of the Turks from Central Europe. As a result of their victories over the Turks the Habsburgs replaced them in Hungary, Croatia, Slavonia, and Transylvania.

In the principality of Transylvania which, in contrast to Hungary, maintained its autonomy, the establishment of Austrian rule was accomplished through negotiations which lasted from 1685 to 1691. Transylvania's constitutional status within the Habsburg Empire was determined by the Leopoldine Diploma of December 1691 according to which the political system based on the three privileged nations and the four recognized religions was maintained, the country was to be known as a great principality which retained possession also of the districts comprising the *Partium* and whose ruler (Prince) was the emperor himself who would govern through governors. The center of governance, and the seat of the Diet, was to be Alba Iulia. It was later moved to Sibiu and, after 1790, to Cluj.

The Banat, which fell under Habsburg rule only in 1718, was governed militarily until 1751 at which time civilian governance took over in all parts except in the military frontier districts.

To ensure direct control over the Transylvanian government by Vienna, a Transylvanian Court Chancellery was established in Vienna in 1694. The new institution was separate from that of Hungary. Similarly, the financial administration and military units of Transylvania were also subordinated to Vienna. For the same reasons of state Catholicism, the official religion of the Habsburgs, was given political powers designed to ensure its preponderance over the Protestant denominations which outnumbered the Catholics among the recognized religions, and was also aided in its proselytizing among Protestants and even Romanian Orthodox, who were regarded as heretics "temporarily" tolerated in Transylvania.

The Romanians, who were excluded from public and political life also by the provisions of the Leopoldine Diploma of 1691, were to be tempted from as early as 1692 by the possibility of securing equality of rights and treatment were they to join or unite with the Roman Catholic Church. The initial requirement was recognition of the validity of the decisions of the Council of Florence of 1439 that had tried to unite the Eastern and Western rites. A consequential improvement of social and political standards for the Romanians of Transylvania was, of course, implicit. Nevertheless, the Union was accepted only in several stages. An initial debate occurred in 1697 at the synod of Alba Iulia which was called by the Orthodox metropolitan of Transylvania, Teofil. His successor, the metropolitan Atanasie Anghel, and 28 archpriests signed the Union in October 1698. Emperor Leopold I confirmed it on 16 February 1699 through a special Leopoldine diploma. A second union diploma followed in March 1701 on the occasion of the reannointing of the formerly Orthodox metropolitan as Uniate bishop. By this second diploma, the Emperor granted the equality of rights promised to the clergy also to laymen who accepted the union *(quinque etiam Saeculares, et plebeae conditionis homines)*.[1] These promises, however, were never fulfilled because the diploma was rejected by the Transylvanian nobles' Diet on the grounds that it was in contradiction to the fundamental provisions of the constitutional diploma of 1691.

Faced with the Diet's fierce opposition, based on apprehension of the threat posed to the class interests of the three privileged nations, the Emperor, engaged in the War of Spanish Succession, abandoned his promises. Even though the imperial decision could

---

[1] Kurt Wessely, "A doua diplomă leopoldină" (The Second Leopoldine Diploma), *Analele Academiei Române. Mem. Secţ. Ist.*, Series III, 1939, XX, p. 288, Plate VII.

be ignored, for the Romanians the document was an eye opener since it showed them that their position could be viewed also in a different light from that codified by abusive nobiliary legislation. The imperial diploma was to become important in the Romanian political movement which was to be initiated by Bishop Inochentie Micu Clain (1728–1751).

VIII

THE DEMOGRAPHY OF TRANSYLVANIA
IN THE MIDDLE AGES AND IN MODERN TIMES

During the early centuries of the Middle Ages there were no censuses or statistics.[1] Demographic data calculated on the basis of parish and fiscal registers are incomplete and approximate.

There are other sources, however, which can indirectly provide demographic and statistical information. When used with care and critically they provide important references. Thus, the thirteenth-century chronicler Simon of Kéza, in his account of the origins and settlement of the Szeklers, states that they were not living in Pannonia but, together with the Romanians, in the mountains on Pannonia's border and also that, being mixed with the Romanians, they learned the Romanians' writing *(Non tamen in plano Pannoniae, sed cum Blackis in montibus confinii sortem habuerunt, unde Blackis commixti literis ipsorum uti perhibentur)*.[2] The severe measures against the Orthodox Romanians of the Banat and Transylvania adopted in the fourteenth century by the Catholic Angevin kings (see pp. 15—17) also testify to the presence in large numbers, and for a long time by then, of Romanians in those regions.

By the sixteenth century one may speak of demographic appraisals and ample and direct statistics derived from records of local humanists and foreign travelers in Hungary and Transylvania. The most widely read and most referred to work in its times and later is that of Georg

---

[1] Ștefan Pascu, *Voievodatul Transilvaniei* (The Transylvanian Voevodeship) Cluj-Napoca, 1979, vol. II, pp. 325 ff.

[2] Simon de Kéza, "Gesta Hungarorum," in *Scriptores rerum Hungaricarum tempore ducum regumque stirpis Arpadianae gestarum*. Szentpétery Emericus Ed., Budapest, vol. I, 1937, pp. 162—163.

Reicherstorffer of Sibiu, first secretary and counsellor to the court in Buda and later employed by Ferdinand I of Habsburg. In his *Chorographia Transilvaniae* (1550), Reicherstorffer states that the Romanians live throughout the land, on all estates and in all villages.[3] His contemporary, Bishop Antonius Verancsics, states that the number of Romanians easily equals that of all Szeklers, Magyars, and Saxons together.[4] The Italian Giovanandrea Gromo calls the Banat *Valachia Cisalpina* or *Valachia Citeriore* — in juxtaposition to Wallachia which was *Valachia Transalpina* — because of the Romanian demographic preponderance.[5] As for the Romanians of northwestern Transylvania the fact mentioned by the French traveler J. Bongars, in 1585, that during his travels there he heard more Romanian spoken than Hungarian,[6] is significant also with respect to the numbers of those speaking the Romanian language. At the same time, in 1584, the Jesuit Antonio Posevino, after recording the fact, in his work on *Transilvania*, that the Romanians were living all over the land stresses the existence of many Romanians in Szekler regions *(sono però con essi /Ciculi/ misti molti Valachi)*.[7] The evidence thus shows the presence of Romanians — representing a homogeneous half of the population in juxtaposition to the other heterogeneous half consisting of Magyars, Saxons, and Szeklers — throughout Transylvania.

In the seventeenth century, the conclusions drawn by the Saxon scholar Johann Tröster (1660) and the observations of the German traveler, the pastor Conrad Jacob Hildebrandt (1656–1658) agree in that they believed that there were more Romanians than all other inhabitants put together.[8]

[3] G. Reicherstorffer, *op. cit.*, Vindobona, p. 28.
[4] A. Verancsics, "De situ Transylvaniae, Moldaviae et Transalpinae," in Verancsics Antal, *Összes munkái*. Pest, 1857, vol. I, p. 143.
[5] Giovanandrea Gromo, "Compendio di tutto il regno posedutto dal re Giovanni Transilvano ...," *Apulum*, II (1943–45), pp. 165–166 and 154–156.
[6] Hurmuzaki-Iorga, *Documente*, vol. XI, p. 190.
[7] A. Veress, *Fontes Rerum Transylvanicarum*, III (1913), pp. 43, 52–53.
[8] J. Tröster, *Das alte und neue Teutsche Dacia*. Nürnberg, 1600, pp. 322, 388; Franz Babinger, Ed., *Conrad Jacob Hildebrandt's "Dreifache schwedische Gesandtschaftreise nach Siebenbürgen, der Ukraine und Constantinopel."* Leiden, 1937, pp. 20–21.

The overwhelmingly Romanian character of the population of Transylvania was also recognized by Vienna. Empress Maria Theresa stated in her letter of 28 March 1748 addressed to the Bishop of Munkács, Manuel Olsavsky, that she was sending him on a religious mission *in dicto Principatu nostro Valachico*[9] or, in other words, "to our Romanian principality, Transylvania." A memorandum of the year 1750, regarding the dual character — religious and political — of Catholic proselytizing among Romanians, states expressly that Transylvania is almost entirely Romanian and that the same Romanians inhabit, together with the Ruthenians, a large part of Hungary *(Transsilvania fere ex integro Valachica est; iidem Valachi cum Ruthenis Hungariae partem bonam incolunt)*.[10]

That information is confirmed, both in words and figures, by the Szekler scholar Benkö Joseph in his work on "The Great Principality of Transylvania." "The number of Romanians", writes Benkö, "is so large that it not only equals but actually exceeds by far that of other peoples in Transylvania" *(Tantus namque est numerus Valachorum, ut reliquorum omnium Transylvaniae populorum personas non modo adaequent sed et multo superent)*.[11] Numerically, the demographic situation between 1761 and 1768 — exclusive of the Bîrsa territory — reveals the existence of 547,243 Romanians and 392,190 members of other nationalities. (According to religious denominations the distribution is as follows: 140,043 Magyar and Szekler Calvinists; 130,365 Saxon Lutherans; 93,135 Magyar, Szekler, and a few Saxon Catholics; 28,647 Socinians or Unitarians who are grouped together with Greeks — actually Aromanians who migrated from south of the Danube — Jews, Gypsies, and others).[12]

Another statistical source related only to Romanians, prepared between 1760 and 1762 on orders of the commanding general of Transylvania, Adolf von Buccow, and which includes also the Bîrsa territory — unrecorded in Benkö's figure — show a total of 155,434

---

[9] *Orientalia Christiana Periodica*. Roma, 1959, vol. XXV, 1—2, pp. 58—59.
[10] Mathias Bernath, *Habsburg und die Anfänge der rumänischen Nationsbildung*, Leiden, 1972, p. 55, note 2.
[11] Joseph Benkö, *Transsylvania sive Magnus Transsylvaniae Principatus olim Dacia Mediterranea dictus*. Vindobona, 1778, vol. I, p. 472.
[12] *Ibid*.

Romanian families. When multiplied by five (father, mother, and three children) the total number of Romanians would be 777,170.[13]

In the "General Description of Transylvania" presented to the future emperor Joseph II by Baron von Preiss, the commanding general of the Principality, for Joseph's edification prior to his voyage to Transylvania in 1773, the following figures — related to the total population of 1,066,017 inhabitants — are given:[14]

Uniates (119,232) and Non-Uniates (558,076): Romanians	677,308 (63.5%)
Lutherans: Saxons	130,884 (12.3%)
Catholics, Calvinists, and Unitarians: Magyars and Szeklers	257,825 (24.2%)

These figures, which indicate a Romanian majority of over 288,500 above that of all other nationalities combined, still do not reveal the actual numbers of Romanians in the Habsburg Empire as they do not include the Banat and the counties of Crişana which were under separate administration.

As for the statistics of the nineteenth century it is important to differentiate between the two distinct phases of administrative rule in Transylvania, i.e. the period of administrative and political autonomy prior to 1867 and that of Magyar governance from Budapest, within the framework of the dual Austro-Hungarian monarchy after the *Ausgleich* of that year.

For the period antedating 1867 the data are as follows:

According to J. Söllner, for the period before 1846, the number of Romanians was 1,290,970; that of Magyars and Szeklers, 616,009;

---

[13] Virgil Ciobanu "Statistica românilor ardeleni din anii 1760—1762" (Statistics of the Romanians of Transylvania of the Years 1760—1762), *Anuarul Institutului de Istorie Naţională*, Cluj, III (1924—25), pp. 616—700.

[14] C. Sassu, *Românii şi ungurii. Premize istorice* (Romanians and Hungarians. Historical Premises). Bucureşti, 1940, Annex 6; Bernath, *Habsburg und die Anfänge*, pp. 220 ff.

and that of Saxons 214,133.[15] For 1846, J. Hain shows 1,369,911 Romanians; 566,760 Magyars and Szeklers; 250,000 Saxons.[16] For 1850, E. A. Bielz shows 1,227,276 Romanians; 536,011 Magyars and Szeklers; 192,482 Saxons.[17] For the same year, 1850, Hunfalvy Paul gives these figures: 1,220,901 Romanians; 535,544 Magyars and Szeklers; 192,218 Saxons.[18] For 1856 Hunfalvy's figures are 1,286,039 Romanians; 555,064 Magyars and Szeklers; 193,774 Saxons.[19]

During that period, the total number of Romanians inhabiting Hungary, the Banat, and Transylvania amounted in 1846, according to J. Hain, to 2,499,866; in 1856, according to Hunfalvy P., to 2,598,591; and in 1869, according to A. Ficker, to 2,647,200.[20]

Despite the discrepancies in the figures for the same year provided by one or another author, and taking into account also the loss of lives which occurred in the revolutionary years 1848—1849, the above statistical data reveal a constant increase in the population of all nationalities—Romanians, Magyars and Szeklers, and Saxons — with a Romanian majority prevailing at all times.

The situation was to change during the Magyar administration of the dualist Austro-Hungarian period between 1867 and 1918. The official statistics record a significant decline in the number of Romanians and of non-Magyar people in general together with a steep increase of the Magyar population. A comparison of the totals for the years from 1846 to 1900 speaks for itself. On the eve of the revolution, according to the statistical data provided by the Magyar statistician Fényes Elek, the population of the Hungarian

---

[15] J. Söllner, *Statistik des Grossfürstentums Siebenbürgen.* Hermannstadt, 1856, pp. 297 ff.

[16] J. Hain, *Handbuch der Statistik des österreichischen Kaiserstaates.* Wien, 1852, vol. I, pp. 211—212.

[17] E. A. Bielz, *Handbuch der Landeskunde Siebenbürgens. Eine physikalisch-statistisch-topographische Beschreibung dieses Landes.* Hermannstadt, 1857, p. 143.

[18] Hunfalvy Paul, *Ethnographie von Ungarn.* Budapest, 1877, pp. 359—360.

[19] *Ibid.*

[20] Dr. A. Ficker, *Die Völkerstämme der Oesterreichisch-ungarischen Monarchie, ihre Gebiete, Gränzen u. Inseln.* Wien, 1869, p. 90.

kingdom and of the Transylvanian principality amounted to a total of 14,120,000 inhabitants.[21] Of these, some 5,250,000 were Magyars. In 1900, the official number of Magyars reached 8,679,014 which represents a spectacular increment of over 3,000,000 people.[22] The Romanian population apparently increased by a mere 285,400 inhabitants, from 2,499,860 in 1846 to 2,785,265 in 1900.

The policy of Magyarization was clearly telling; still, despite efforts to attain a Magyar majority within the confines of the kingdom the Magyar percentage could not exceed 45.4 percent.

In 1915, the German publicist Friederick Naumann[23] declared that "The Magyars know that per capita they represent less than half the population." And Naumann continues: "Hence it results that those who, through their search for national domination, have created the state are forced on the one hand to reject — overtly or covertly — egalitarian democracy and, on the other, to increase the number of Magyars." And he concludes: "Thus, Magyar sovereignty in the state cannot be secured until the Magyars will amount to over 50% of the population. Herein lies the significance of their anxiety to Magyarize. In fact, meaningful results are recorded in Magyar statistics. Indeed, between 1900 and 1910, the number of those who declared that their mother tongue was Magyar, or of those who could not oppose their being recorded as speaking that language, increased from 45.4% to 48.9%, and of those who declared that they knew Magyar from 52.9% to 57.4%. Should matters continue like this for another few decades a certain majority will be secured in statistics and, later, in actuality."

---

[21] Fényes Elek, *Magyarország leírása* (A Description of Hungary). Pest, 1847, vol. I, pp. 19 and 25.
[22] R. W. Seton-Watson (Scotus Viator), *Racial Problems in Hungary*. London, 1908, pp. 3—4.
[23] Fr. Naumann, *Mitteleuropa*. Berlin, 1919, pp. 89—90.

IX

THE RESISTANCE OF THE TRANSYLVANIAN ROMANIANS TO THE POLICIES OF THE MAGYAR NOBILITY AND OF THE HABSBURG REGIME IN THE SIXTEENTH, SEVENTEENTH, AND EIGHTEENTH CENTURIES

Whenever nobiliary oppression caused violent reactions by the serfs, the Romanians fought side-by-side with Magyar and Szekler serfs in anti-aristocratic social movements. That was true of the fourteenth and fifteenth centuries (see p.18). Later too, in 1514, in the peasant war of Hungary and Transylvania led by the Szekler George Dozsa, the rebel army counted many a Romanian in its ranks and among its leaders, such as the voevode of Ciuci (of the Mureş region), lesser nobles from Maramureş, and others. The consequence of the war was the increase in feudal oppression, voted by the Diet which met at Pest in November 1514 and codified in the famous *Tripartitum*, or code in three parts, prepared by the legal adviser Werböczy István. The peasantry was reduced to total and permanent servitude *(mera et perpetua servitute et rusticitate)*. Also in the sixteenth century, in 1569, the anti-Magyar and anti-Turkish rebellion led by the Romanian Gheorghe Crăciun had 10,000 peasants, mostly from the Sălaj and Satu Mare regions, in arms. At the beginning of the eighteenth century the Romanians of Maramureş, Satu Mare, Bihor, Arad, and all the way to the eastern border of Transylvania seeking liberation from feudal servitude joined the rebellion led by Francis Rákóczi II against the Habsburgs.

Resistance was also manifested by rejection of attempts at religious conversion. In the sixteenth and seventeenth centuries Lutheran and Calvinist preachers sought, albeit unsuccessfully, to convert Orthodox Romanians. As stated by the Patriarch of Constantinople Cyril Lukaris, in 1629, the failure of those attempts should be attributed to the blood and sentimental ties between the Romanians of Transylvania and those of Wallachia and Moldavia, which provided the moral support for resistance to Protestant entreaties.[1] The Romanians of Wallachia and Moldavia, according to the same source, supported their Transylvanian brethren "if not with weapons at least with secret advice."[2]

The permanence of Transylvanian-Moldavian-Wallachian ties existed also in cultural-religious matters. Thus, for instance, during the second half of the sixteenth century Coresi's printing shop — which was moved from Tîrgoviște, in Wallachia, to Brașov — added to the commercial fame of the Transylvanian city also that of being a center for dissemination of Romanian culture. The books printed by Coresi for the benefit of all Romanians with contributions from all parts inhabited by Romanians were of decisive importance for the formation of the Romanian literary language. It should also be remembered that in Transylvania, in the seventeenth century, at Alba Iulia the work of the learned Orthodox Metropolitan Simeon Ștefan equaled that of Metropolitan Varlaam of Moldavia both in the defense of the faith and in the development of the Romanian language.

The strengthening of Romanian Orthodoxy reached its heights in the last decades of the seventeenth century in Wallachia and Moldavia which, in turn, strengthened religious ties with Transylvania and stimulated the resistance of Transylvanian Romanians to the attempts to subordinate them to the Catholic Church. There are three main periods during which Romanian opposition to Union with the Roman Catholic Church was most marked. During the first such period, from 1697 until 1711, support from the other side of the Carpathians was characterized by the reaction of the Orthodox hierarchy and of the princes, in Wallachia and Moldavia. Toward the middle of that century, from 1744 to 1746, the opposition assumed the form of a true war of religious independence within the Hune-

---

[1] The Latin text in *Török-Magyarkori Allam-Okmánytár* (State Documents from the Turkish-Magyar Epoch). Pest, 1869, vol. II, pp. 137—140; I. Lupaș, *Documente istorice transilvane* (Transylvanian Historic Documents). Cluj, 1940, vol. I, p. 178 (Translation from the original Latin).

[2] *Ibid.*

doara, Sibiu, Făgăraș, and Brașov regions. Later, between 1757 and 1761, the agitation spread to the Apuseni mountains as far north as Maramureș. A decree of toleration, ambiguous from its inception, granted by Maria Theresa in July 1759 aggravated rather than calmed the situation. Only the terror unleashed by the military commander of the principality, General von Buccow, including the burning or tearing down of monasteries as well as mass arrests and convictions, led to the quashing of the rebellions. However, these actions together with the revolutionary experience gained by the masses paved the way to the great uprising which was to break out in 1784.

The eighteenth century also recorded the beginning of the political struggle of Transylvania's Romanians to win recognition as a nation. The opening salvo, fired by Inochentie Micu Clain (1692—1768) a bishop from 1728 until 1751, was based on the Leopoldine promises related to religious union, particularly those contained in the second diploma which had been rejected by the Diet. From its inception, the struggle assumed national dimensions. Bishop Inochentie based the struggle on the interests of the entire people and assigned priority to the national over the religious factors. In other words, he asked that the Romanians be given equal rights with members of the other Transylvanian "nations," that the Romanian masses be regarded as equal with the masses of the other "nations," that they be granted the right of free movement, access to education and crafts, et cetera. His demands, recorded in numerous petitions, were based both on natural and historic rights of his people. He invoked, on the one hand, the overwhelming number of, and the overwhelming public and military tasks assigned to, Romanians in Transylvania and, on the other, the superiority of the origin, the antiquity and continuity of the Romanian people on Transylvanian soil. "Nothing should be decided about us, without us and in our absence"[3] was the basis of his revendications.

The political objectives and arguments formulated by Inochentie are the foundations upon which was to be erected the entire national struggle of the Romanians of Transylvania which aimed at

[3] Prodan, *Supplex*, p. 136.

the general economic, social, political, and cultural elevation of the people.

The early followers of Inochentie's program were young Uniate priests who had been educated in Catholic seminaries in Hungary and, particularly, in Vienna and Rome. Their close contacts with Central and Western European cultural movements allowed these young men to study history, which enriched their knowledge of the origins of their people, philology, which provided scientific data on the Latin character of the Romanian language, natural law, and the philosophy of the Enlightenment. Thus was asserted the Transylvanian School, the ideological and cultural movement of Romanian intellectuals from Transylvania at the end of the eighteenth and beginning of the nineteenth centuries. Its leaders included the linguist Samuil Micu and the historians Gheorghe Șincai and Petru Maior who contributed to the crystallization and formulation in the all-important *Supplex Libellus Valachorum* of 1791 (see pp. 42—43) of the postulates advanced and defended by Inochentie Micu Clain during the tenure of his bishopric for sixteen years and even after his removal from that office in later years.

The philological, historic, and literary work of the Transylvanian School played a major role not only in affirming the cultural and national rights of the Transylvanian Romanians but also in the ideological preparation of the entire Romanian national movement of the nineteenth century.

X

HOREA'S UPRISING
OF 1784 AND ITS HISTORIC SIGNIFICANCE

The Romanians of Transylvania hoped that the assumption of power, in 1780, by Joseph II — an "Enlightened Despot" — would result in a lessening of the rigors of imperial administration. Joseph II visited Transylvania on two occasions, in 1773 and in 1783, and was acquainted with the situation of the Romanians as witnessed by his own words:

> "These poor Wallachian subjects, who undoubtedly are the oldest and most numerous inhabitants of Transylvania, are nevertheless so abused and overwhelmed by unjust actions from all, Hungarians, or Saxons, that we must recognize that their fate is worthy of pity, and it is surprising that there are still so many of them and that they did not flee." [1]

During his voyage in the Banat and Transylvania of 1773, the Romanian serfs could often hear words of encouragement, spoken in their own language by the young coruler in reply to their numerous complaints. On the other hand, the nobility and the privileged became alarmed when he spoke of his intended reforms as he believed that the condition of the Romanians could not remain as it was *(So kann es nicht verbleiben, wie es jetzt ist)*. [2]

In simple terms, then, it may be said that hope was to encourage the oppressed in seeking rights either through petitions or even through uprisings, whereas concern and alarm over the possibility

---
[1] Bernath, *Habsburg und die Anfänge*, p. 220.
[2] *Ibid.*, pp. 209—220, 225.

of their having to share their "privileges" with the tolerated was to strengthen the determination of the privileged to oppose, by all means, the revendications of the Romanians. This accounts also for their reinterpretation of past history, denying the Romanians' autochthonism and advancing the tendentious theory of Romanian immigration, at a later date, from south of the Danube.

The social and national oppression of the Romanian people in Transylvania was underlined also — as shown by Mathias Bernath — in the various reports drafted for Joseph II in preparation of his proposed visit in 1773. In "Description of the Province," signed by General von Preiss, it is stated repeatedly that the Romanians are deprived of all constitutional rights although they are the most numerous and oldest inhabitants and despite their select ancestry. They are not regarded as a "nation" but only as a tolerated "people" (populum).[3] Another report states explicitly that "the nobility considers the Romanian as a man destined to be a slave, a man whose happiness must be derived from his being allowed to breathe the air."[4]

The great uprising began in 1784 in the Apuseni mountains; it was led by Horea, Cloşca and Crişan and lasted two months, November—December. It started in the Zarand *comitatus* from where it spread rapidly to the neighboring Hunedoara and Alba counties and, hence, southward toward the Banat, Sibiu and Făgăraş, northward as far as Maramureş and eastward toward Cluj. In fact, the crisis encompassed all of Transylvania and even beyond. The rebels' program, as stated on 11 November, called for the dissolution of the nobility, for every nobleman to live off his job and pay taxes just as the rest of the ordinary people, for division of the land of the nobility; all-in-all then, a revolutionary program which, if enacted, would have brought about the collapse of the feudal order. It was the first time that a peasant uprising recorded so radical a program — going beyond the agrarian claims of the revolutions in the subsequent century.

[3] *Ibid.*, pp. 140, 192—193.
[4] *Ibid.*, p. 142.

The imperial army, heeding the appeal of the Magyar nobility which had been attacked by the serfs, put down the rebellion in blood. Its leaders, Horea ānd Cloşca, were broken on the wheel, on 28 February 1785, before thousands of Romanian peasants who were brought to witness the punishment of their leaders. Parts of the bodies of the two leaders were exhibited in the regions where the serfs rose up in arms.

The echo of the uprising was heard throughout Europe. Calendars, pamphlets, and booklets as well as the press, particularly in France, the German states, the Netherlands, Italy, Spain, and England, contained much information and many commentaries on the causes, course, leaders, and suppression of the rebellion.[5] It goes without saying that diplomatic reports and notes and imperial instructions regarding the uprising were plentiful. In many accounts there were also comments on the past, origins, and living conditions of the Romanian people.

Despite the hostility toward and outright condemnation of the revolt, which reflected particularly the position of Transylvanian ruling circles, it is noteworthy that several expressions of appreciation of the plight of the rebels and of condemnation of the oppressive nobility also appeared. Thus, the German newspaper *Real-Zeitung*, of Erlangen, stressed that the root cause of the rebellion was "the treatment, worse than that of slaves, which the Romanian peasants had to suffer at the hands of the nobility."[6] The future Girondist leader Jacques-Pierre Brissot wrote Emperor Joseph II an "open letter" in which he defended the rights of the Romanian peasants. He stated that in staging the uprising Horea "wanted to free [the Romanian peasants] from servitude" which "legitimized everything." Brissot also described the inhuman living conditions of the Transylvanian Romanians while emphasizing the fact that they represented two-thirds of the population. He concluded by saying that "if the Americans were right in rebelling because they no longer wanted to be taxed without

---

[5] M. Edroiu, *Răsunetul european al răscoalei lui Horea* (The European Echo of Horea's Uprising). Cluj-Napoca, 1979, p. 231; D. Prodan, *Răscoala lui Horea* (Horea's Uprising). Bucureşti, 1979, vol. II, pp. 674–714.

[6] *Real-Zeitung*, no. 98, 14 December 1784, p. 293.

their consent, the Romanians were all the more so as they had neither property, nor liberty, and were at the mercy of their masters." [7]

Horea's rebellion was also viewed as a prelude to the great changes which were to be brought about, only some five years later, by the French Revolution. "Europe is facing a great change which is still smoldering today" wrote the anonymous author of the pamphlet *Kurze Geschichte der Rebellion in Siebenbürgen...*, which was published in Strasbourg in 1785.[8]

In many of its aspects, however, the rebellion had a national character. First, the serfs imparted a national character to their social struggle through the overwhelming number of Romanian rebels facing the allogeneous minority of their oppressors. Several news items contained in the Western press regarding the course of the rebellion characterize the uprising as such. Both from within the Habsburg Empire and from outside there were imputations that Horea entertained the revival of Dacia with himself as *rex Daciae*. In fact, in the national tradition, the symbol of Dacia as related to Horea's name served to strengthen Romanian solidarity in the quest for the unitary national state.

---

[7] *Seconde lettre d'un défenseur du peuple à l'Empereur Joseph II sur son Règlement concernant l'émigration, et principalement sur la révolte des Valaques*. Dublin, 1785, pp. 70—79. See appendix A, p. 110.

[8] *Kurze Geschichte der Rebellion in Siebenbürgen, ...*, 1785, p. 39.

XI

FROM *SUPPLEX LIBELLUS VALACHORUM* (1791) TO THE REVOLUTION OF 1848

The memorandum of March 1791, known as *Supplex Libellus Valachorum*, marked the height of the national-political struggle of Transylvania's Romanians at the end of the eighteenth century. The *Supplex* comprises in a comprehensive manner the demands severally stated in the petitions presented by Inochentie Micu Clain between 1728 and 1751 and of later date. The memorandum was a collective effort, the work of high-level Romanian intellectuals in the second half of the eighteenth century. It contained the signatures of "the Clergy, the Nobility, the Military and the Urban Estates of the whole Romanian nation in Transylvania." Long in the making, the *Supplex* of 1791 falls within the historical framework of revolutionary revendications then pervasive in Europe and was reflective also of the political unrest manifest among the peoples of the Habsburg Monarchy following the collapse of Joseph's reforms.

The main goal of the memorandum was the regaining of the ancient rights *(pristina jura)* of which the Romanian nation had been deprived altogether. The census of 1787, mentioned in the memorandum, showed that the Romanians numbered "almost a whole million" of the total of "one million and some seven hundred thousand" inhabitants of the principality. Therefore, it was asked that the demeaning and undignified appellations of *tolerated, admitted*, and the like be publicly abolished and that the Romanian nation's civil and political rights be fully restored; that the equality of rights of the entire nation (clergy, nobility, and ordinary folk) with those of other nations of Transylvania be proclaimed; that the nation be granted proportional representation, to the size of the population, in public life. Moreover, the suppliants demanded that in the event that their grievances should not be considered by the Diet, the Romanian nation be given the right to meet in a "national assembly" (or national congress) of their own which would elect deputies or representatives who would defend its interests whenever necessary.

The lengthy and thoroughly documented demands were based on the historic rights of the Romanian people, on natural equity and on the principles of civil society, on the rights of man and citizen in general. The Roman origins of the Romanians, the length of their inhabitation and their continuity on the territory of former Dacia were stressed. However, emphasis was also placed on the Romanians' rights derived, by necessity, from their overwhelming numerical superiority in Transylvania.

Under these circumstances, then, the Romanians' constitutional demands, if agreed to, would have resulted in granting the Romanian nation not only the status of "fourth nation" in Transylvania but also that of dominant nation. As could be expected, the memorandum was angrily rejected by the Diet of the three privileged nations. Nevertheless, the opposition of the nobility and the restrictions of the Habsburg regime, no matter how influential they might have been, could no longer arrest the development of the Romanian nation and the setting up of its national state.

With the nineteenth century the movement of emancipation of the Romanian people entered, throughout the territories inhabited by Romanians, the road of fulfilment of national unity and independence. The champions of the new Transylvanian generation which followed the Transylvanian School and the *Supplex Libellus Valachorum* virtually abolished political frontiers by blending efforts with people on the other side of the Carpathians for the attainment of a common national goal. Often, the same scholars worked alternately on one or another side of the Carpathians. The presence of Transylvanian intellectuals in the Romanian principalities, where they stayed years on end as professors, together with the frequent voyages of Moldavian and Wallachian counterparts to the Banat and Transylvania, made fundamental contributions to the circulation and exchange of new ideas.

So unitary did Romanian national ideology become by the third decade of the nineteenth century that the transition from ideology to practice, in the succession of revolutionary manifestations which occurred in the Banat, Transylvania, Wallachia, and Moldavia before 1848, concern and collaboration for achievement of the political unity of all Romanians within the confines of a unitary national state, were evidenced with equal conviction and strength in all Romanian lands. It is easy to understand why such revolutionary actions could not occur concomitantly and in the same manner in Transylvania, Moldavia, and Wallachia but alternated among the three provinces. Internal and external factors caused the movement of the forces of preparation and manifestation thereof from one province to another and, on occasion, even beyond their frontiers. Still, at all times, participation and assistance of all Romanian lands was envisaged.

Thus, plans for attaining a Romanian national state, with emphasis on programming, were most actively conceived in the Banat and Transylvania in 1834—1835. In those years, as stated also by the Magyar historian Kovács Endre, "the Daco-Romanian idea, whose practical goal was the plan to unite the Romanians from both sides of the Carpathians into a single homeland" was most vigorously advocated there.[1] Characteristic too were the principles of organization of the "unitary Romanian republic" which included equality of rights and obligations, opportunity for every citizen to hold any public function, universal suffrage, abolition of all titles of nobility, emancipation of the peasantry, land reform, free education, equality of religious cults, and so forth.

The goal and program of the unsuccessful attempt of 1834—1835 were to be found, in the years immediately following, also in the plans and revolutionary agitations of Wallachia and then again in the Banat, in the movement initiated by Eftimie Murgu. In 1845—1846, revolutionary planning was more active in Moldavia. A period of activity abroad ensued; on the eve of the revolution of 1848 it was equally intense both within and outside the borders. It was Nicolae Bălcescu who launched a radical revolutionary program which sought the achievement of the unitary Romanian state, demo-

---

[1] Kovács Endre, *A lengyel kérdés a reformkori Magyarországon* (The Polish Problem in the Reform Era in Hungary). Budapest, 1959.

cratically organized. On 1 January 1847 Bălcescu stressed that "our goal cannot be any other than national unity of all Romanians." Revealing, too, is the motto which he launched also at that time: "Romanianism, thus, is our flag, under which we must summon all Romanians."[2]

The prospects of development of Romanianism as a consequence of the maturing of the national consciousness on both sides of the Carpathians were regarded with concern by the Magyar nobility, the holder of political power in Transylvania. Therefore, the process of liberalization and democratization of the feudal order in Transylvania was slow and painful. Inasmuch as liberalization and democratization entailed, in the first place, the granting of rights to the Romanian majority such action would have been politically suicidal for the Magyar nobility. To avoid the inevitable Romanianizing of Transylvania's political structure the nobility resorted to the palliative solution of forcible Magyarization. The cue was given from as early as 1833 by the most influential representative of Magyardom in Transylvania, Wesselényi Miklós, who wrote: "I regard it not only correct but also very necessary that the common people not enjoy national and representation rights unless they were to become truly Magyar, if they were to become part of the nation whose rights they wish to benefit from, if they were to identify themselves with that nation by language and customs. That is not a means of discrimination or constraint. For such a wonderful reward the price to pay is only the learning of a language, which is not too high for anyone. Thus, those who do not now speak Hungarian or are not Magyars would gradually, as they are Magyarized, receive national rights also."[3] Emancipation thus meant Magyarization.

Later, in 1842, the Diet adopted the draft of the law which provided the gradual adoption, over a period of ten years, of the Magyar language as Transylvania's official language. Vehement protests from Romanians and Saxons ensued against that "shameful deal and unjust

---

[2] N. Bălcescu, *Privire asupra stării de față, asupra trecutului și viitorului patriei noastre* (Survey of the Present Status, of the Past and Future of our Fatherland). C. Bodea edition. Bălcești pe Topolog, 1971, p. 40.

[3] Wesselényi Miklós, *Balitélekről* (On Prejudices). București, 1833, pp. 232—233.

law" — as it was called by the Romanian scholar and fighter for political rights S. Bărnuțiu — or against "the sentence of destruction of Romanian nationality" — as it was called by G. Barițiu, another Romanian political leader. In his article of that year, "Der Sprachkampf in Siebenbürgen," on the struggle for the primacy of the language of the state in Transylvania, St. L. Roth clearly asserted: "I do not see a need for imposing an official language. We already have a language of the land. It is not the German language, nor is it the Magyar, but it is the Romanian! No matter how we, the nations represented in the Diet, may twist or turn there is nothing we can do about that. That is reality." And, with premonition, the Saxon pastor warned the Magyar nobility "For they have sown the wind and they shall reap the whirlwind." [4]

The ultimate reasons for the policy of Magyarization so perseverently pursued by the nobility are best and most revealingly stated by Wesselényi himself in 1846. His arguments are clearly and unequivocally stated in a letter to Louis Kossuth:

> "We must not forget that our nationality now exists through this nobility. On all scales of values and of reason our nobility has sufficient defects and not a few sins; but the truth is that Magyardom exists only or almost only within its ranks. If through ruination of thousands of nobles their number were to become insignificant, most of the millions who would replace them would not be Magyars. It would be a totally abnormal situation and a mathematical absurdity for the small number to be and to weigh more than the large one. That absurdity of the nationality and language of the Magyars — so much fewer in numbers — being above the masses — much larger — of the other peoples has been possible and is possible only by the fact that with us the great majority $= 0$ and only the number of the nobility has any value. Since the majority of the nobility is obviously Hungarian it is for that reason, and

---

[4] St. L. Roth, *Der Sprachkampf in Siebenbürgen. Eine Beleuchtung des Woher und Wohin*. Kronstadt, 1842, pp. 47–48. See Appendix C, p. 138.

for none other, that it is not absurd that its language and its people should be above the others.

But that will change should the non-Magyar millions constitute the majority, numerically and juridically. Our present aristocracy may disappear; no great harm, as it is not worth much. Another one will replace it on the basis of abilities, money, or estate which, to be sure, will be larger than the present one. It could be, and it is reasonable to believe, that it will be better, more intelligent. But it will not be Magyar! These are circumstances which require great attention and which make a revolution even more disastrous for us than for other nations. The French Revolution chased the aristocracy but another one was formed, also French because it was made up of Frenchmen. In Galicia, the Polish nobility was slaughtered by the Polish peasantry; the place once occupied by the dead will be either empty, or will be occupied by descendants of the murderers, but all will remain Poles. It is otherwise with us..."[5]

Thus, Wesselényi acknowledged the numerical minority of Magyars in Transylvania and the fact that Magyardom was represented primarily by a few thousand nobles among millions of non-Magyars. It was therefore that he suggested, in 1833, the Magyarization of the "common people" who were non-Magyar. Therefore, too, in 1846 he insisted that the nobility, such as it was, bad rather than good, had to be retained and treated with kid gloves for, should it disappear, it would be replaced only by the non-Magyar millions which would have meant, as a matter of course, the collapse of Magyar political domination.

The Magyar aristocracy regarded the union of Transylvania to Hungary, too, as a way out of the dangerous demographic and political imbalance since union would bring about the consolidation of its position on both counts. The numerical imbalance and the social and political antagonism between these two worlds, that of the

---

[5] Z. Ferenczi, "Kossuth és Wesselényi s az ürbérügye 1846—1847-ben" (Kossuth and Wesselényi and the Urbarial Question), *Századok*, 36(1902), pp. 53—56, 140—146. See also Cornelia Bodea, *1848 la români* (1848 with the Romanians), București, 1982, pp. 302—305.

thousands of nobles or privileged Magyars and that of the millions of oppressed non-Magyars, were confirmed by the radical Pest paper of the revolutionary years, *Marczius Tizenötödike*. More than that, recognition of that situation is related expressly to the problem of Transylvania's annexation to Hungary. The above-mentioned paper, in its issue of 26 May 1848, includes the following statement: "In Transylvania two forces *(két hatalom)* will decide the fate of the union — the Diet and the Romanian people. The Diet represents only a few hundred men, the Romanians represent Transylvania as a whole ... The Union of Transylvania to Hungary without the Romanians' consent is something we should not embark on."[6]

It should be noted that whereas in 1848 the program of the Magyar revolution, which included the incorporation of Transylvania into Hungary, was drawn up within the confines of the nobles' Diet of Pozsony and approved by the other nobles' Diet of Cluj, the program of the Romanian revolution was acknowledged and approved by the assembly of the entire Romanian population of Transylvania. That difference was noted also by the mentioned Pest paper. At the Blaj assembly of the Romanians of 15 May 1848 — that is ten days before the appearance of the issue of *Marczius Tizenötödike* mentioned above — Simion Bărnuțiu, the ideologue of the revolution of the Transylvanian Romanians, stated in his speech that "The true liberty of any nation can only be national," while the masses expressed their desire to "be united with the country" which meant the Romanian principalities from beyond the Carpathians.

Actually, national Romanian policy was based on two fundamental principles, to wit, the liberty and independence of the Romanian nation, and indictment and rejection of the union of Transylvania to Hungary which was determined by the Magyar nobility without the Romanians' consent and against the Romanians' interests. On the basis of the right of self-determination of peoples, the Romanians sought national equality, not in the feudal sense of securing privileges but in that of securing the new rights, the freedoms consecrated in almost every country by the European revolution. Speaking through Simeon Bărnuțiu, the leaders of the Romanian revolution

[6] *Marczius Tizenötödike*, 26 May 1848.

clarified their position on 15 May 1848 as follows: "The Romanian nation informs the coinhabiting nations that in its desire to form and organize itself on a national basis it entertains no evil thought against other nations and recognizes the same rights for all nations and wants to respect those rights sincerely while seeking mutual respect based on justice. Therefore, the Romanian nation neither wishes to rule over other nations nor will it accept to be subject to others, but wants equal rights for all."[7]

As for the Romanians' rejection of the union of Transylvania to Hungary their attitude became all-the-more intransigent as it became all-the-more evident that the union was made solely for the purpose of ensuring Magyar supremacy. On 25 September 1848, at the third people's assembly of Blaj, the people's representatives stated solemnly that the Romanian people "does not wish to recognize the union of Transylvania to Hungary; rather, it asks for a Transylvanian Diet and a provisional government for Transylvania, both comprising Romanian, Magyar, German representatives, numerically proportional to the size of each nation, to decide "the future order of the country!"[8]

"Nature has placed us in a country where we would sweat together in cultivating it, where together we would taste the sweetness of its fruit," said Avram Iancu, the military leader of the Romanian revolution, while the militant revolutionary Simonffy Jozsef stated that: "Romanians and Magyars are in urgent need of close brotherhood. Only thus will they assure their existence. Therefore, he who sows discord, hatred and causes bloodshed between these two sister-nations is a traitor to his own nation."[9]

It was also Avram Iancu who said that the failure to recognize their nationality and the despotic measures taken by the Magyar

---

[7] S. Bărnuțiu, "Raporturile românilor cu ungurii și principiile libertății naționale" (The Romanians' Relations with the Hungarians and the Principles of National Freedom) in G. Bogdan-Duică, *Românii și Ungurii* (The Romanians and the Hungarians), Cluj, 1924, paragraph LXVIII. See also C. Bodea, *1848 la români*, pp. 478—479.

[8] T. V. Păcățian, *Cartea de aur* (The Golden Book), vol. I, 1904, pp. 427 ff.

[9] Silviu Dragomir, *Avram Iancu*. București, 1968. Second Edition, p. 242.

aristocracy against them made the Romanians rise up in arms. And "what is striking"—comments the American historian Keith Hitchins — "is Iancu's use of the term 'brothers' when addressing himself to the Magyars after almost nine months of hard battles."[10]

"For true liberty and recognition of the political existence of the Romanian nation we live and die" writes, in turn, Ioan Buteanu — one of the martyrs and commanders of the Romanians' revolution — to major Csutak of the Magyar army.[11]

Let us also mention the conclusion drawn by the American historian and publicist Francis Bowen in his article "The War of Race in Hungary":[12] "The Magyars, indeed, fought with great gallantry; it was hardly possible to avoid sympathizing with a people who struggled so bravely against immense odds. But their cause was bad; they sought to defend their antiquated feudal institutions, and their unjust and excessive privileges as an order and a race, against the incursion of the liberal ideas and the reformatory spirit of the nineteenth century."

In the heat of battle gross accusations were uttered and acts of cruelty were committed by both sides. But even then attempts were made to find a common language for overcoming difficulties, both by the Romanian and by the Magyar side. Some were even made by Romanians from the other side of the Carpathians. "The Romanian revolutionaries [from the Principalities] truly wanted to join the Magyar struggle against Vienna," writes Kovács Endre. But such collaboration was impeded by factors which were not justly recognized. For, in the words of the same Magyar historian, "Bălcescu, Golescu-Negru, and other radical leaders of the revolution, who placed the abolition of serfdom at the top of their program, extended also to the Transylvanian Romanians their plans for social and national

[10] Keith Hitchins, *Orthodoxy and Nationality. Andrei Şaguna and the Rumanians of Transylvania, 1846—1873*. Cambridge, Mass., 1977, p. 76.

[11] Ion Ghica, *Amintiri din pribegia după 1848* (Recollections from the Exile after 1848), Bucureşti, 1940. Also Deák Imre, *1848. A szabadságharc története levelekben*. (1848. History of the Struggle for Liberty in Letters), Budapest, 1942, p. 382—383..

[12] *North American Review*, Boston, Mass., January 1850.

liberation; they could not remain indifferent toward the fate of the Romanian peasant of Transylvania which—as has been since recognized also in Magyar historiography — did not improve materially even after the spring of 1848, and for which reason the movements of the Romanian peasantry of Transylvania continued. The indecisiveness shown by the Magyar government in the agrarian problem led to exacerbation of the status of the nationalities which, in addition to social oppression, had to suffer also national oppression."[13]

---

[13] Kovács Endre, *A Kossuth emigrácio es az europai szabadságmozgalmak* (The Kossuth Emigration and the European Liberation Movements). Budapest, 1967, p. 275.

XII

THE AUSTRO-HUNGARIAN COMPROMISE OF 1867 AND ITS IMPACT ON TRANSYLVANIA

After the repression of the Revolution of 1848—1849 the system of absolute governance was reintroduced throughout the Habsburg Monarchy. It lasted from 1850 until 1860. In Transylvania, now directly subordinated to Vienna, devoid of its Diet and Court Chancellery, social and national antagonisms pursued their course as they were encouraged, on the one hand, by the representatives of the Habsburgs and, on the other, by the nobility which sought to regain its lost position of leadership.

National agitations spread throughout the Empire and territories around it. By the end of the fifties two capital events took place which had broad implications for Austrian internal and foreign policies. One was the establishment of the Romanian unitary national state on the Empire's southeastern frontier through the union of Moldavia and Wallachia in January 1859; the other was the removal from the Empire of the Italian lands of Lombardy, Tuscany, and Modena following the Peace of Zurich of November 1859. The first swayed the center of gravity of Romanian unification toward the Romania on the other side of the Carpathians, while the second marked the beginning of the dismemberment of the multinational Habsburg conglomerate and the first stage of the unification of Italy.

As a result of its losing the war with France and Italy, Vienna was forced to abandon its absolutist rule in 1860 and adopt constitutional governance on a federalist basis. The Romanians in the Empire remained separated from a territorial standpoint. Those in the Banat, Crişana, and Maramureş remained under Hungarian administration. Bukovina remained an autonomous duchy as established in 1849. The Transylvanian governance of pre-1848 was restored.

During the brief period of conflict between the interests of Vienna and those of the Magyars, the Romanians of Transylvania — for just as brief a period — saw their hitherto rejected revendications realized. The Diet of Sibiu of 1863—1864, enjoying a primarily Romanian and Saxon majority because of the non-participation of many Magyar deputies in its activities, enacted the "Law of equal national rights" for the Romanian nation and its two religious denominations and recognized the Romanian language as official language of Transylvania.

National equality, however, lasted only until 1865 since negotiations between Vienna and Pest, pursued against the wishes of non-Magyar nationalities, led first to the annulment of the decisions of the democratic Diet of Sibiu in September 1865 and, then, to the reopening of the nobles' Diet at Cluj, in November of the same year, for the "exclusive and sole" purpose of voting the incorporation of Transylvania into Hungary. That was followed by the convocation of the Diet of Pest and, finally, in February 1867, by the official enactment of the *Ausgleich* (Compromise) which established the dualist system of governance in the new Austro-Hungarian Monarchy. The Compromise was thus an agreement or association of two nations against the many *(Unio duarum nationum contra plures)*, but reflecting class interests. "The Magyar [nobles]" — writes Louis Léger — "with their usual egoism thought only of their own interests; [....] they took advantage of their victory to impose heavy domination over Romanians, Serbs, and Slovaks."[1]

Transylvania, once more a ploy of the Habsburgs and the Magyar oligarchy, was placed under direct Hungarian rule. Thus, its autonomy and the laws enacted at Sibiu were done away with. The American historian Robert A. Kann, a scholar well known for his work on the Austro-Hungarian Monarchy, qualifies the Compromise as "one of the most deplorable chapters in the political history of the nationalities of the Monarchy."[2]

---

[1] Louis Léger, *Histoire de l'Autriche-Hongrie*. Paris, 1920, p. 557..
[2] Robert A. Kann, *Das Nationalitätenproblem in der Habsburger Monarchie*, Graz-Köln, 1964, vol. I, p. 315.

The Romanians' reaction against the incorporation of Transylvania into Hungary and, as such, reaction against the Austro-Hungarian Compromise, reached new heights — on both sides of the Carpathians — even as early as the negotiation stage. The Romanian national-political program was made known by the publication, in the press of the Empire and the foreign press, of the memorandum of protest known as the "Blaj Pronunciamento" (15 May 1868). The three principal postulates of that memoradum were (1) Restoration of Transylvania's administrative and political autonomy; (2) Recognition and application of the laws enacted by the Diet of Sibiu of 1863—1864; (3) Reopening of the Transylvanian Diet on the basis of genuine popular representation.

Arthur J. May, another American historian, most knowledgeable of the problems of the Austro-Hungarian Monarchy, emphasizes the fact that in spite of the political union of Transylvania to Hungary, partly because of it, particularist sentiments, even among the Magyar speaking population, continued to be strong. Of all the national minorities in the realm of the Habsburgs — continues the same author — the Romanians, next to the Italians, were the least devoted and the least loyal to the Monarchy.[3]

---

[3] Arthur J. May, *The Habsburg Monarchy, 1867—1914*. New York, 1968, p. 72.

XIII

THE POLICY OF MAGYARIZATION AND NATIONAL OPPRESSION IN THE DUAL MONARCHY

The Budapest government, from the very beginnings of the Austro-Hungarian dualist system, adopted the principle of uninationalism which made the traditional policy of assimilation through Magyarization official. The law granting equal rights to the Romanian nation as well as the other laws enacted by the Diet of Sibiu were abrogated at the time of Transylvania's incorporation into Hungary, in June 1867. These actions placed Transylvania — according to contemporaneous usage — in an "exceptional and discretionary" state, which allowed governance according to the wishes of the government rather than according to law.[1] Legislative enactments differed in their application in Hungary and in Transylvania in that they were enforced in a more discriminatory manner in Transylvania. Thus, for instance, in Transylvania the press law required the depositing of security money for the publishing of any periodical. The required amount could be as high as 10,000 florins. Pre-publication issues had to be submitted to the Ministry of the Interior as well as to the bench and *procuratura* of the appellate court of proper regional jurisdiction. Fines and jail terms up to two years were given for articles deemed contrary to the interests of the state.[2] Likewise, the application of the electoral law, as modified in 1874, differed in Hungary and Transylvania. In Transylvania the application was designed to deny voting rights to Romanians and other subject nationalities

---

[1] G. Bariţiu, *Părţi alese din Istoria Transilvaniei* (Selected Parts From the History of Transylvania). Sibiu, 1891, vol. III pp. 454 ff.

[2] Seton-Watson (Scotus Viator), *Racial Problems*, Appendix X (Roumanian political trials 1886—1896, 1897—1908).

(Slovaks, Serbs, etc.) through higher tax rates on income derived from the land as well as by excluding from voting all those who could not read or write Hungarian. Moreover, illegal districting designed to affect negatively non-Magyar nationalities as well as electoral intimidation and corruption, all contributed to exacerbation of national discrimination.

During the first few years following the Compromise, however, the Budapest government adopted seemingly liberal, moderate, and conciliatory measures in matters related to national problems. The Law of Nationalities or Law XLIV adopted in December 1868, which proclaimed the existence of a sole political nation, to wit "the indivisible unitary Magyar nation," and which decreed that Hungarian be the only official language, did nevertheless envisage the "official usage" *(hivatalos használat)* of the languages of various nationalities under certain circumstances. However, the regulations which determined specific instances and conditions for deviations from the norms were so aleatory that they made for reinterpretations and violations rather than for applications of the letter of the law.

C. A. Macartney, the well-known British historian, accounts for differences between theory and practice in terms of the traditional mentality of the Magyar ruling class. "The Magyar and Magyar-minded ruling class in Hungary had never in history regarded their State as either a-national or multi-national, and they did not do so now. They could not feel that the primacy allowed by the Law to their language was simply a pragmatic concession to administrative efficiency, and that the State was no more 'theirs' than it was the Slovaks' or the Romanians'. They were even convinced — and the conduct of the nationalities in and after 1848 had deepened that conviction — that the very existence of Hungary depended on the maintenance of its Magyar character. While Deák and Eötvös were alive and politically active, their influence was still strong enough to compel a certain moderation, and at least the main provisions of the Law were observed. But the pot was simmering even then, and a big change for the worse set in when the Ministry of the Interior was taken over by Tisza." [3]

The Tisza government lasted from 1875 to 1890. After that no Magyar government looked upon Hungary in any other terms than those viewed from the prism of Magyar supremacy. They differed only as to the depth of Magyarization and the violence of enforcement thereof which ultimately aimed at creating a "unitary Magyar national State, the centre of gravity of the future Magyar-Austrian Monarchy."[4] As the steep downward trend was set by the Tisza government there was no point of turn left to his successors. The unitary Magyar state thus became political dogma. Tisza was violently opposed to universal suffrage claiming that "it would have been even more fatal than any other threat as it would have forever destroyed the [Magyar] national state."[5] Baron Bánffy Dezső, another violent opponent of the non-Magyar nationalities in the Empire[6], while Prime Minister (1895—1899), claimed that "there were no Romanians in Transylvania. In Transylvania there are people of Romanian race, which is altogether different."[7] In 1906, it was also he who stated explicitly that "Hungary's interests demanded adoption of the most extreme Chauvinism," while in 1908 he stated unequivocally that "without Chauvinism nothing can be achieved."[8] Count Albert Appony, in turn, stated ten years before launching his known school law that "inside the Magyar state there can be no nationalities, nor nationality rights. There is only one nationality: every Hungarian subject is Magyar [...]. We do not know a Croat-

---

[3] C. A. Macartney, *The Habsburg Empire, 1790—1918*. New York, 1969, p. 722.

[4] *Ibid*, p. 722. See also I. Tóth Zoltán, "A nemzetiségi kérdés a dualismus korában, 1867—1900" (The Nationalities Question in the Dual Epoch), *Századok*, XC (1956), 3, pp. 379—380.

[5] Horváth Zoltán, *Magyar Századforduló*. Budapest 1967, p. 249.

[6] See *Histoire de la Hongrie, de ses origines à nos jours*, Editions Corvina. Budapest, 1974, pp. 407—411, 605.

[7] Charles Benoist, *Souvenirs*. Paris, 1932, vol. II (1894—1902), p. 367.

[8] R. W. Seton-Watson, *A History of the Roumanians*. London, 1934, pp. 419—420.

ian, or Romanian, or Slovak nationality. And we will never recognize them."[9]

J. Diner-Dénes rightly remarked that "Liberty, equality, independence and nation, liberal, social, national ... these notions have a totally different meaning in Hungary than everywhere else. For the outside world they are, evidently, used in the general meaning, but within the country they are used *consensu omnium* only in a falsified sense."[10]

The area on which laws and measures of Magyarization concentrated with the greatest impact was education. In 1868 public education with a uniform program and curricula prepared by the state was introduced in Transylvania. However, despite the provisions of the Nationalities Law which allowed "free choice of the language of instruction" in conformity with regional demography, not one public elementary school in which the language of instruction was Romanian or Slovak was established. In denominational schools or in those belonging to public or private associations or to individuals, Magyar became part of the curriculum. The Trefort school laws of 1879 and 1883, enacted during Tisza's government, made Magyar the language of instruction in all elementary schools. Non-Magyar teachers were given four years to ready themselves to teach in the Magyar language. The Csáky law of 1891 introduced Magyar also in kindergartens. In 1907, the even more drastic Apponyi Law made Magyar compulsory in all non-Magyar schools attended also by Magyar children. As a consequence, the number of non-Magyar elementary schools decreased by the hundreds. Thus, the number of Romanian elementary schools declined from 2,756 in 1880 to 2,170 by 1914.[11]

National discrimination was evident also in the economic, administrative, political, and cultural areas. Data on functionaries of Romanian origin at the beginning of the twentieth century in Transylvania are self-explanatory. For instance, in higher administration

[9] Benoist, *Souvenirs*, vol. II, p. 375.
[10] J.Diner-Dénes, *La Hongrie, Oligarchie, Nation, Peuple*. Paris, 1927, p. 7.
[11] *Magyar statistikai évkönyv* (Magyar Statistical Yearbook), 1882, Book 9, pp. 94—99 and *Magyarország közoktatás ügye az 1914 évben* (The Problem of Instruction in Hungary in 1914). Budapest, 1917, p. 12.

there were only 5.4 percent Romanians and in counties and towns only 7.4 percent Romanian functionaries. By contrast, Magyar and German public prosecutors and judges amounted to 94.4 percent.[12]

Among the means used to secure Magyarization was also that of Magyarizing people's names. Telkes Simon's work "How to Magyarize Family Names" *(Hogy magyarositsuk a vezetékneveket)* published in Budapest in 1898, provides instructions on how to change non-Magyar names into Magyar ones. To prevent such actions, Romanians would give their children Latin names which could not be changed into Magyar ones.

Resistance to national oppression was evident in all fields and assumed the most varied forms. Since 1869, to implement the policy of non-acceptance of the *status quo* for Transylvania, the Romanian National Party adopted "passivism", that is boycotting the Pest Parliament, as a political weapon. As previously mentioned, the leaders of the Romanians of Hungary and the Banat — themselves organized in a Romanian National Party — sought to collaborate within the framework of Parliamentary procedure for a period of eight years. In 1881 they joined with the Transylvanians in the formation of a sole Romanian National Party of Transylvania and Hungary, and continued parliamentary "activism" until 1887.

Passivism and activism, both practiced until 1887, the total passivism observed between 1887 and 1905 as well as the activism again pursued after 1905, were supplemented by manifestations affecting all aspects of political, economic, and cultural life. All were designed to safeguard Romanian national identity, to keep the national problem at the forefront of public concern both within and outside the empire.

A decisive factor in stimulating resistance against Magyarization and denationalization was the magnetic attraction of the Romanian national state on the other side of the Carpathians. From 1848 until 1918 the desire of "unification with the country" — or with the Romanians on the other side of the Carpathians — which was

---

[12] *Magyar statisztikai közlemények* (Magyar Statistical Communications). Budapest, 1906, New series, vol XVI, pp. 136—145.

expressed at Blaj (see p. 48) gained momentum gradually, reaching peaks of crisis from time to time.

Even Louis Kossuth stated in 1850 that the idea of the unitary "Daco-Roman" state had to preoccupy the Romanians as "there is no human force which could prevent the sentiment, desire, and instinct from remaining alive in the peoples' conscience."[13] In 1859, when Alexandru Ioan Cuza was elected Prince of Moldavia and Wallachia alike, accomplishing the union of the two Romanian principalities as an expression of the national will, "the enthusiasm of the Romanians in Transylvania was perhaps greater than in the two principalities."[14] That statement, by the Transylvanian 1848 revolutionary Papiu-Ilarian, was neither isolated, nor circumstantial. This is tellingly confirmed by the fact that in many parts of the empire the Romanians, both intellectuals and common people, greeted one another in those days with the words "long live uncle Ioan,"[15] a direct reference to the prince of the United Principalities, Alexandru Ioan Cuza, the new symbol of national unity.

Even during a period of political relaxation in Transylvania, such as that of the years 1863—1864, it was easy to observe in the Romanian's minds "a latent expectation that one day, by union, a numerous and mighty Roumanian nation will be formed."[16] In the years antedating the *Ausgleich* of 1867, collaboration

---

[13] From the letter of Kossuth to Teleki László, 18 September 1850, for transmittal to the Romanian politician I. Heliade Rădulescu. (Magyar Nemzeti Muzeum, Budapest), quoted by Cornelia Bodea in *Din Istoria Transilvaniei* (From Transylvania's History), second edition. Bucharest, 1963, p. 175.

[14] See Al. Papiu-Ilarian, "Memorand despre raporturile românilor cu nemții, cu slavii și cu ungurii în timp de pace și în cazul unei revoluțiuni în răsăritul Europei, prezentat principelui Cuza în 1860" (Memorandum on the Romanians' relations with the Germans, Slavs and Hungarians in Time of Peace and in the Case of a Revolution in Eastern Europe, Submitted to Prince Cuza in 1860), *Revista pentru Istorie, Archeologie și Filologie*, I, 1883, pp. 134—146.

[15] In a letter addressed from Buda (Pest) on December 4, 1860 by Alexandru Roman, official provincial translator to the editorial office of the paper *Naționalul* in Bucharest (Library of the Academy of the Socialist Republic of Romania, Ms. 1668, ff. 382—383), quoted by C. Bodea in *Din Istoria Transilvaniei*, p. 187.

[16] Charles Boner, *Transylvania, Its Products and Its People*. London, 1865, p. 394.

between the Romanians of the Monarchy and those on the other side of the Carpathians included even plans for military revolt.[17] During the Romanian war of independence of 1877—1878, in addition to the moral and material aid given by Transylvanian Romanians, former border guards at the Romanian frontier as well as in other parts of Transylvania declared themselves ready to enlist for action on the Russo-Romanian front in Bulgaria. Only the opposition of Russia's foreign minister Gorchakov, bound by the secret Reichstadt agreement of 1876 with Austro-Hungary, made for the absence of units of Transylvanian volunteers. Individuals participated however.

The winning and international recognition of Romania's independence and then the establishment of the kingdom, in 1881, stimulated both the reorganization and intensification of the political life of the Romanians in the Austro-Hungarian Monarchy. In 1881 these Romanians decided to resume, on a large scale, the policy of presenting protests and memoranda against Magyar hegemony to Europe's public opinion. The series of such actions, begun in 1882, included the memorandum of 1887 and culminated in the memorandist movement of 1892—1894, which ended with the celebrated trial held in Cluj and the conviction of the authors of the Memorandum of 1892. The European press followed these issues closely.

Close relations between the Romanians on both side of the Carpathians were evident throughout those years. Arthur Nicolson, the British Consul General in Budapest between 1888 and 1892, one of the keenest observers of the national resistance movement developing in Hungary in those years, wrote, for instance: "During the tour which I recently made in Transylvania, I found among all intelligent classes of Roumans a feeling of widespread discontent, grounded on more serious and specific grievances than those which exist among the Serb and Croat nationalities, and to which expression and form has been given by a well-organized system of opposition. Furthermore, and this is peculiar to the situation in Transylvania, there is not only openly expressed sympathy in Roumania

---

[17] Hitchins, *Orthodoxy and Nationality*, p. 163.

with the grievances of their brethren, but means have of late been adopted in that country also towards assisting in their alleviation."[18]

For that matter, in 1889 the Saxon paper of Sibiu, *Siebenbürgisch-Deutsches Tageblatt*, stated unequivocally that the very close and direct, secular, relations made "these two ethnic entities, though politically separate, to feel as one national entity."[19]

For the Saxons themselves as well as for their economic interests — as may be seen also in the secret report submitted by F. Lachmann to the Austro-Hungarian authorities in Vienna in 1880 — the union of Transylvania and Romania seemed inevitable. In Lachmann's words, "I had the opportunity to talk to several members of the Saxon colony who did not hesitate to tell me that their only hope lies in the province [Transylvania] becoming Romanian someday."[20]

It was also from Saxon circles that consul Nicolson learned of the strong national solidarity characterizing all Romanian social strata and classes in Transylvania. The information gathered from his "long conversations with moderate, sensible Saxons" — especially Bishop Teutsch — unanimously highlighted the futility of the Magyarizing policy: "In isolated instances, and among some of the weaker-kneed races, this policy may have occasional successes, but they are convinced that efforts to turn a Rouman into a Magyar must terminate in failure"[21].

Faced with such centrifugal tendencies the Austro-Hungarian cabinet in Vienna, in order to counteract the support provided by the empire's neighboring national states, adopted the policy of overt or secret alliances with Italy, Serbia, and Romania. That policy, however, was proven useless by the natural course of events.

---

[18] *Report on the Political Situation in Transylvania.* See Appendix, D, p. 144.
[19] "Rumänien und Oesterreich-Ungarn", *Siebenbürgisch-Deutsches Tageblatt*, 9 October 1889.
[20] Ernst R. V. Rutkovski "Oesterreich-Ungarn und Rumänien 1880—1883. Die Proclamierung des Königsreiches und die rumänische Irredente", *Südost-Forschungen*, XXV (1965), pp. 204—205 (Report Lachmann, 9 August 1880).
[21] See Appendix D, p. 159.

XIV

THE UNION OF TRANSYLVANIA WITH ROMANIA

As it is known, the political crisis of the Austro-Hungarian monarchy was aggravated by the military defeats suffered during World War I and by economic calamities, all of which favored attainment of the goals of the nationalities of the empire. In the fall of 1918 Austria-Hungary disintegrated.

According to the official census of 1910, in the area of today's Transylvania,[1] Romanians represented 53.8% of its total population of 5,248,540 inhabitants, Magyars 28.6%, and Germans 10.2%.[2]

At a meeting held in Oradea on 12 October 1918, the Executive Committee of the Romanian National Party adopted a declaration of principles which affirmed the nations' right to self-determination and the willingness of Transylvania's Romanians to use it. The contents of the declaration, read by the Romanian deputy Al. Vaida-Voevod at the meeting of the Budapest Parliament on 18 October was as follows:

"Vis-a-vis the situation created by the world war the Executive Committee of the Romanian National Party of Transylvania and Hungary finds that the consequences of the war justify the century-long claims of this [Romanian] nation for full national freedom. On the basis of the natural right whereby

[1] To wit, the old Principality of Transylvania, the Banat, Crişana and Maramureş — the last two being designated as *Partium* in the past.
[2] The figures have been calculated on the basis of the census of 1910. See *A Magyar korona országainak 1910 evi népszámlalása* (Census of the countries belonging to the Magyar Crown) II. rész. Budapest, 1912. For the adjacent counties which were partially incorporated into Transylvania, the figures are based on the population of each commune. See Dr. Sabin Manuilă, "Aspects démographiques de la Transylvanie," in *La Transylvanie. Ouvrage publié par l'Institut d'histoire nationale de Cluj*. Bucarest, 1938, pp. 798 and 822.

every nation may alone and freely decide its fate — a right now recognized also by the Magyar government through the Monarchy's armistice proposal — the Romanian nation now wishes to make use of that right and, therefore, claims also for itself the right to alone determine its place among free nations, without any external interference. The national organ of the Romanian nation of Hungary and Transylvania does not recognize the right of this parliament and of this government to regard themselves as representatives of the Romanian nation in order to act as representatives of the interests of the Romanian nation of Hungary and Transylvania at the general peace congress. This, because the defense of its rights can be entrusted by the Romanian nation only to spokesmen named by its own national assembly. The Romanian nation which lives in the Austro-Hungarian Monarchy expects and demands, after century-long sufferings, recognition and fulfilment of its incontrovertible and inalienable rights to full national existence." [3]

The Romanian Central National Council (established in Budapest on 31 October 1918, and moved subsequently to Arad) decided on 10 November to assume "total power of governance over the territories inhabited by Romanians in Transylvania and Hungary." Consequently, military guards and Romanian national councils were formed throughout Transylvania, the Banat, Crișana, and Maramureș.

In early November 1918 one hundred representatives of Magyar intellectual circles in Budapest — among whom the well-known Ady Endre, Kodály Zoltán, and Bartók Béla — issued a manifesto in favor of the creation of the independent national Hungarian state and of the liberation of the oppressed peoples in the Austro-Hungarian Monarchy. However, the signatories of the Manifesto advocated the maintenance of the eastern half of the old Austria-

[3] *Gazeta poporului*, Sibiu, I, 1918, no. 43, October 27. *Marea Unire de la 1 decembrie 1918* (The Great Union of 1 December 1918). București, 1943, pp. 24—25.

Hungary through a federalization of the nationalities within that territory. This in fact annulled the self-determination principle itself by preventing those nationalities from uniting with their own nations. Such a federation, that amounted to an attempt at saving former Hungary, was not bound to take shape, as could be inferred from the manifesto itself: "Old Hungary has collapsed [...] Our purposes do no longer run counter to the purposes of others. We have thereby put an end to war. We don't want to fight any longer. We have no further claims against sister nations. We, too, consider ourselves a renewed nation, a now liberated force, just like the brethren who happily are beginning a new life out of the ruins of the monarchy. We awaken with a conscience relieved by the fact that we are no longer forced to be the pillars of oppression." And the manifesto read in conclusion: "Let us live side by side in peace as free nations with other free nations." [4]

In Transylvania, at Tîrgu Mureș, the Magyar National Council, established on 31 October 1918, worked for "the right of self-determination of non-Magyar peoples of Hungary" and for "recognition of all established or about to be established national states" as stated by the Council's president Antalffy Endre. [5]

Negotiations held at Arad, between 13 and 15 November 1918, between representatives of the Magyar National Council and the delegation of the Magyar government, led by Oszkár Jászi, on the one hand, and the delegation of the Romanian National Council, on the other, were broken off without any achievement. The Magyar delegation, while recognizing the right of the Romanian nation to freely decide its own fate on the basis of Wilsonian principles, sought to retain Magyar sovereignty in Transylvania on the basis of the concept of "integrity of the Magyar state." The Romanian National Council stated unflinchingly that "the Romanian nation rightly

---

[4] Apud Jászi Oszkár, "Amintiri despre tratativele mele din Arad cu Comitetul național român" (Recollections of my Negotiations in Arad with the Romanian National Council), in *Gazeta Ardealului*, Cluj, I, 1921, no. 236, December 11 (translated from the Magyar original in the *Napkelet* review, Cluj).

[5] See László Bányai, *Destin commun, traditions fraternelles*, Bucharest, 1972, pp. 102—103 (Bibliotheca Historica Romaniae. Études).

seeks total statal independence and cannot allow the obscuring of that right by provisional solutions." [6]

Transylvania, together with the Banat, Crişana, and Maramureş, were united with Romania on 1 December 1918 by the freely expressed decision of the over 100,000 representatives of the Romanian population of those lands who constituted the Great Assembly of Alba Iulia. The resolution adopted at that time proclaimed that "the National Assembly of all Romanians of Transylvania, the Banat, and Hungary assembled through their rightful representatives at Alba Iulia on 18 November (1 December) 1918, decrees the union of these Romanians and of all territories inhabited by them with Romania."

The decision of Alba Iulia was freely taken, without any interference from the outside. The Assembly met on land occupied neither by the forces of the Entente nor by the troops commanded by General Mackensen. The Romanian army itself was deployed on Transylvania's borders, at a considerable distance from Alba Iulia, on 1 December 1918. In the period before the meeting of the Assembly as well as during the actual sessions, public order in Transylvania was maintained exclusively by the national guards. The Assembly of Alba Iulia acted in accordance with the principle of self-determination of nations regarded by Woodrow Wilson as the fundamental criterion for a new and equitable world order. The Proclamation of Alba Iulia, based on justice and respect for the rights of coinhabiting peoples, provided for equal political and religious rights for all coinhabiting peoples in Transylvania, for universal, direct, equal, and secret ballot, for radical land reform, for workers' rights comparable to those enjoyed in the most advanced industrial states, for religious freedom, for freedom of the press, and for elimination of war as the means for solving international problems.

As it was easy to foresee, the exponents of the old Austro-Hungarian regime could not hail the act of December 1, 1918 of Transylvania's union with Romania. But the newly constituted Magyar National Council in Tîrgu Mureş cooperated from the very beginning with Consiliul Dirigent Român (the Romanian Ruling Council). [7]

---

[6] *Românul*, VII, 1918, no. 7.

[7] *Brève histoire de la Transylvanie*, Bucharest, 1965, p. 425.

The Szekler deputy, baron Dr. J. Fay, in his statement in the Bucharest Parliament on 19 February 1920 asserted his nationality's confidence as regards the union with Romania, and voiced its dissociation from every act running counter to the union. [8]

As a matter of fact, in an official note addressed to the Romanian, the Czechoslovak, and the Yugoslav governments on 30 April 1919, Béla Kun, then Hungarian foreign minister, showed that the previous Hungarian government had rejected the proposals of these states "by invoking the so-called historic right, which derived from old oppression the right to continue oppression. We discarded that principle from the very day we assumed power and have stated repeatedly and solemnly that we do not rely on the principle of territorial integrity and we are now informing you directly that we recognize unequivocally your territorial claims." [9]

Strong approval for the union came also from the inhabitants of German descent in Transylvania and the Banat. The Saxon Central Council *(Sächsischer Zentralausschus)* and the Saxon National Council *(Sächsischer Nationalrat)* meeting in the Assembly of Mediaş of 8 November 1919 adopted a resolution in which it was stated: "The Union of Transylvania and of the areas in Hungary populated by Romanians with Romania has brought forth a compact territory, whose unity is based on its ethnographic realities. In view of these facts and persuaded as we are of the global significance of that act, the Saxon people of Transylvania, in accordance with the principle of self-determination, pronounces itself in favor of the union of Transylvania with Romania, conveys its fraternal salute to the Romanian people and cordially congratulates it upon achievement of its national ideals." [10] The Saxon people takes cognizance not

---

[8] *Desbaterile Adunării Deputaţilor* (Bucureşti), 21 Febr. 1920, no. 41, p. 594.

[9] *A Magyar munkásmozgalom törtenetek valogattot dokumentumai* (Select Documents of the Hungarian Workers' Movement). Budapest, 1960, vol. 6/a. p. 355.

[10] A. Armbruster und H. F. Jaeger, "Über die Stellung der Deutschen Bevölkerung Siebenbürgens und des Banats zur Vereinigung von 1918", *Revue Roumaine d'Histoire*, VII, 1968/6, p. 1087—1097; also C. Göllner. "Die Stellungnahme der Siebenbürger Sachsen zur Vereinigung Transsilvaniens mit Rumänien", *Forschungen zur Volks- und Landeskunde*, IX, 1966/2, p. 29—38.

only of a historic process of global significance but also of the legitimate right of the Romanian people to unite and establish a state.

In turn, the Banat Swabians opted for union with Romania at their assembly of 10 August 1919 and on 13 August sent a memorandum to the Paris Peace Conference which asserted: "By voting in favor of the resolution for union with Romania, the National Assembly of the Swabian people wanted to emphasize [...] the categoric desire unanimously expressed by the Assembly that the entire Swabian people be reunited with the Romanian people [...]. Centuries of life together have taught us to appreciate our neighbors and coinhabitants for what they really are, and our recent experience has only strengthened our conviction that only union with Romania would provide us with sufficient guarantees for our existence and our progress."[11]

However, even before the beginning of the Paris Peace Conference on 18 January 1919, Transylvania's union with Romania had been carried out on the basis of the right of self-determination which the Romanians had exercised in the same manner as the other peoples of the former Austro-Hungarian Monarchy. The Hungarian delegation to the Conference adopted the position of retention of "the integrity of the Hungarian state," in other words, it claimed all territories the populations of which had chosen to unite with Romania, Serbia, and Czechoslovakia. The representatives of Great Britain, France, Italy, and the United States to the Conference studied, through their commissions and experts, the voluminous files relevant to Hungarian claims for some two months. It should be noted that these commissions, chaired by political leaders such as André Tardieu, the future prime-minister of France, or Lord Curzon, included no representatives from Romania, Serbia, and Czechoslovakia which countries, although victorious in World War I, were regarded by the Conference as litigants just as defeated Hungary.

Upon examination of the Magyar documentation the Conference, on 6 May 1920, handed to Hungary's representatives President

---

[11] "Memoriul prezentat la Conferința de pace de către delegația șvabilor din Banat" (Memorandum Presented at the Peace Conference by the Delegation of the Swabians from the Banat), *Revista Institutului social Banat-Crișana*, XII (1943), p. 420.

Millerand's letter [12] which contained the answer to the Hungarian notes as well as the final text of the Peace Treaty of Trianon. In this letter it was stated that "the allied and associated powers studied with the greatest care the notes by which the Hungarian delegation expressed its views on the peace terms which had been communicated to it," but that "upon profound consideration they took the decision not to change any point of the territorial clauses contained in the peace terms. They decided in that manner because they were convinced that any modification of frontiers fixed by them would involve graver drawbacks than those denounced by the Hungarian delegation."

The decision of the allied powers was based on careful study of the ethnic situation in Central Europe. Recently, an Austrian historian writing on the Paris Conference showed that "until that time, in modern history, there was never a peace conference prepared for such a long time and so thoroughly, never before were all controversial issues examined and debated, following depositions by interested parties, so thoroughly." [13]

The political leaders, diplomats, and experts, however, reached the same conclusion as those who had sought solutions to the national problems of that part of the world. President Millerand's letter pointed out that "the ethnographic conditions of Central Europe are such that, indeed, it is impossible for the whole length of political frontiers to coincide with ethnic frontiers. It follows that certain nuclei of Magyar population — and the allied and associated powers resigned themselves not without regrets to that necessity — will find themselves under the sovereignty of another state. Yet it cannot

---

[12] "Ce n'est pas sans de mûres réflexions que les puissances alliées et associées ont pris le parti de ne modifier sur aucun point les clauses territoriales contenues dans les conditions de paix. Si elle s'y sont résolues, c'est parce qu'elles se sont convaincues que toutes les modifications des frontières fixées par elles entraîneraient de plus graves inconvénients que ceux que dénonce la délégation hongroise." Arhiva M.A.E., București (Archives of the Ministry of Foreign Affairs) Holding 71/1920—1944, *Hungary*, vol. 1, p. 1—5.

[13] Fritz Fellner, "Die Friedensordnung von Paris 1919—1920," *XVe Congrès International des Sciences Historiques. Rapports.* Bucarest, 1980 vol. I, p. 221.

be claimed, because of this situation, that it would have been better to leave the old territorial order unchanged. The continuation of an order, though millenary, is unjustified when it has been shown that it is contrary to justice."

The participants in the Paris Peace Conference were aware of the fact that the allied powers had only to confirm, at the international level, the validity of already accomplished historic acts. This is revealed by the same document: "The wish of the peoples was expressed in October and November 1918, when the Dual Monarchy was collapsing, in the union of the long-oppressed people with their Italian, Romanian, Yugoslav, or Czechoslovak brothers. Events since then are further testimonials to the sentiments entertained by the nationalities once subject to the Crown of Saint Stephen. The belated measures of the Hungarian government designed to satisfy the nationalities' quest for autonomy no longer hold out illusions; they in no way affect the essential truth which is that over many a year all Magyar political efforts were aimed at stifling the voice of ethnic minorities."

On 4 June 1920 the treaty of peace was signed in the palace of Trianon. According to the British historian R. W. Seton-Watson, "The Treaty of Trianon ends the most momentous epoch in the whole history of the Roumanian race." [14]

---

[14] Seton-Watson, *History of the Roumanians*, p. 554. "The most momentous epoch," because it was one of fight for a great ideal, as the British diplomat A.W.A. Leeper also singled out as early as in 1918: "The Union does not mean Transylvania's annexation to Romania or Romania's annexation to Transylvania, as asserted by people fond of paradoxes; the Union means the building of a united nation formed of the population of Romania and of the provinces in the Austro-Hungarian Monarchy, it means fulfilment of the ideal of the Romanians in Austria-Hungary." See *La Roumanie*. Paris, I, 1918, no. 30, August 8.

XV

THE CONSEQUENCES OF THE UNION OF TRANSYLVANIA WITH ROMANIA

The act of 1 December 1918 at Alba Iulia reinstated an important part of the Romanian people in their normal human rights and set Transylvania in the natural unity of Romania, from a political, ethnical and economic standpoint.

From the union of 1918 until 1940, when the northwestern part of Transylvania was given to Hungary by the Vienna Diktat, Transylvania showed continuous progress despite the ravages of World War I and the great economic crisis of 1929—1933.[1]

As a result of the land reform, carried out throughout Romania in 1921, land was given to 310,583 peasants in Transylvania, Romanian, Magyar, Szekler, or Saxon, and the percentage of small landholdings — less than 10 hectares — increased from 34.1% to 56.5%. The production of cereals and of farm animals doubled between 1922 and 1940. The Magyar government sought to prevent the carrying out of the land reform in Romania as it sought exemption from expropriation of lands belonging to big Hungarian landowners in Transylvania who had opted for Hungarian citizenship. This meant in fact disregard of Romania's sovereignty. The Romanian position was defended before the Council of the League of Nations by Romania's foreign minister, Nicolae Titulescu. Numerous well-known specialists in international law, Belgian, Swiss, French, German, Italian, and Dutch, rendered opinions on the "optants" problem

---

[1] The figures below are from Silviu Dragomir, *La Transylvanie avant et après l'Arbitrage de Vienne*. Sibiu, 1943, pp. 30—35.

which emphasized the fundamental legality of Romania's actions.[2] The litigation ended, after more than eight years of debates, at the International Court of Justice of the Hague through a Romanian-Hungarian agreement which fully recognized the validity of the expropriation principle.

The number of commercial and of small industrial enterprises increased in Transylvania between 1918 and 1940 by 61.6% from 37,115 to 96,661. As far as big industry was concerned the number of enterprises increased in the period 1919—1938 from 1,161 — with 81,589 employees — to 1,691 — with 129,603 employees. Significant progress was recorded in technology and urban development, the highway system was expanded by 1,069 kilometers, 2,348 new bridges were built and 5,182 old ones were modernized, the railway network was expanded to encompass previously isolated areas of Transylvania. The health system showed significant progress as attested by the fact that, whereas in 1918 there was one doctor per 8,375 inhabitants in towns and 11,599 in villages, in 1940 there was one physician per 5,183 inhabitants in towns and 6,770 in villages. Sixty-nine general hospitals were built and 837 new dispensaries were established. In state hospitals alone the number of beds increased from 6,731 to 13,437. Mortality rates decreased from $26^0/_{00}$ in 1914 to $18^0/_{00}$ in 1940. 2,553 new school buildings were erected. In 1918 Transylvania had 3,727 elementary schools; that number rose by 1936 to 5,641. During that interval the number of teachers increased from 8,962 to 14,455. Before 1918 Transylvania had one elementary school per 1,400 inhabitants; by 1935, there was one such school per 900 inhabitants. Under Magyar rule there was one high school per 560,000 Romanians; after the union there was one high school per 20,000 Magyars.

The Romanian Constitution of 1923 adopted the principle of equality of rights for all ethnic groups. The law of 1924 on education provided that in localities where languages other than Romanian

---

[2] See *La Réforme agraire en Roumanie et les Optants hongrois de Transylvanie devant la Société des Nations*. Paris, 1927, pp. 588.

were also spoken instruction be also given in the respective languages. The Magyar-language press flourished. György Lajos, a specialist in the history of Magyar literature, showed, on the basis of an analysis of the status of Hungarian newspapers and weeklies published in Romania between 1919 and 1923, that "under the new government the number of newspapers and weeklies has doubled." He also showed that "the same astounding leap forward is true also of religious and educational papers as it is of professional publications. The number of those in the first category trebled after 1919 while that of the second increased eight fold."[3]

---

[3] György Lajos, "A Romániai magyar idöszaki sajto öt esztendeje (1919—1923)" (Five Years of Magyar Press in Romania), in *Erdélyi Írodalmi Szemle*, (1924), no. 7.

XVI

MAGYAR REVISIONISM BETWEEN THE TWO WORLD WARS

The Magyar government, even though it signed and applied the provisions of the treaty of 4 June 1920, did not abandon its aspirations to reconstitute the kingdom of Saint Stephen. Encouraged by Mussolini's revisionism, Hungary concluded an alliance with Italy on 5 April 1927. Fàscist Italy and Horthy's Hungary thus initiated officially their policy of revisionism directed against the member states of the Little Entente — Czechoslovakia, Yugoslavia, and Romania — and France. The promoters of Magyar revisionism, both within and outside of Hungary, were, in the words of R. W. Seton-Watson, "oligarchs whose powers were based on great estates, entail and mortmain, open ballots, control of the press and restriction of association and assembly." [1]

In his lectures given in England, in November 1933, Count Bethlen Istvan, while recognizing the impossibility of annexing Transylvania to Hungary, still demanded its "autonomy" for no other purpose than to raise an apple of discord in Central Europe. All lectures and all revisionist propaganda contained not one argument which had not been presented by Hungary to, and rejected by, the 1920 Peace Conference. Horthy's revisionism, as well as the actions of Nazi Germany and Fascist Italy, violated the Covenant of the League of Nations which required all member nations to respect the territorial integrity and current political independence of all members of the League.

In response to the Magyar revisionist campaign, which expressly revealed Hungary's desire to increase its territory at the expense of that of its neighbors, Nicolae Titulescu, Romania's foreign minister

---

[1] R. W. Seton-Watson, *Treaty Revision and the Hungarian Frontiers*. London, 1934, p. 18.

and twice president of the League of Nations, stated in Romania's Parliament, on 4 April 1934: "We have told Hungary repeatedly and we are telling her again today: a mutual understanding is in the general interest of peace and in the interest of the two countries. But an understanding becomes impossible if it raises the question of changing borders. Hungary knows just as well as I do that through its propaganda it will not be able to alter our borders by one square meter. Then why does she do it? The explanation is simple. The aim of revisionist manifestations is to demoralize the countries against which they are directed and to dangle before Hungarian public opinion an uncertain hope of a better time to come."[2] And he continued: "The present borders of Central Europe reflect the century-long evolution of an idea of justice. The question of the borders of Central Europe is a closed issue and not an issue which must be raised again. To any attempt to start anew we will answer: the issue has been settled."[3]

Fascism in Hungary became more pronounced during the regime of Gömbös Gyula (1932—1935). The withdrawal of Hitler's Germany from the Disarmament Conference and from the League of Nations in October 1933 marked the beginning of its expansionist policy. The muddy international situation resulting from Nazi Germany's denunciation of the Treaty of Locarno (7 March 1936) and reoccupation of the Saar region, the revisionist speech of Mussolini in support of Horthy's Hungary delivered in Milano on 1 November 1936, the signing of the anti-Comintern pact by Germany and Japan, all influenced the internal and foreign policies of Horthy's Hungary. Through the annexation of Austria in March 1938 Germany became Hungary's neighbor. Admiral Horthy's pompous visit to Germany, in August 1937, led to the first concrete manifestations of Hungarian revisionism. The Munich Conference and Agreement (29—30 September 1938) whereby Czechoslovakia had to cede the Sudetenland to Germany were followed by the first arbitral award of Vienna of 2 November 1938 whereby Hungary occupied a territory of 12,000 square kilometers, inhabited by some 1,000,000

---

[2] Nicolae Titulescu, *Discursuri* (Speeches), București, 1967, p. 409.
[3] *Ibid*, p. 417.

people, in southern Slovakia. Romania, faithful to its international agreements, opposed the dismemberment of Czechoslovakia. In October 1938, the Romanian government rejected the proposal of the Polish government that Romania participate in the territorial mutilation of Czechoslovakia and, at the same time, the government viewed with favor the possibility of allowing the Soviet airforce to fly over Romanian territory in order to assist Czechoslovakia. In March 1939, concurrently with the invasion of Czechoslovakia by Nazi forces, Hungary occupied another part of Slovakia.

On 11 April 1939, Horthy's Hungary withdrew from the League of Nations at a time when Magyar revisionism was the main German instrument for exerting pressure on Romania and Yugoslavia.

XVII

THE VIENNA DIKTAT AND ITS SIGNIFICANCE

Horthy's revisionism was crowned with success on 30 August 1940 when in the Belvedere Palace in Vienna the foreign ministers of the Third Reich and of Italy, von Ribbentrop and Ciano, imposed on Romania the Diktat whereby northwestern Transylvania was to be annexed to Hungary. The Vienna Diktat occurred at a time when most of Europe had either fallen to or was dominated by the Axis powers and when totally-isolated Romania was at the mercy of Nazi Germany and Fascist Italy. As early as 15 July 1940 Hitler wrote to King Carol II in no uncertain terms that Romania had to accept territorial losses to its neighbors and that "every attempt to overcome the dangers menacing your country by tactical maneuvers of any sort whatsoever must and will fail. The outcome sooner or later—and perhaps in a very short time—may even be the destruction of Rumania."[1]

The occupation of that part of Romanian land had been decided on earlier by the fascist powers and their Hungarian ally. At Vienna it was only necessary to communicate to the Romanian representatives the arbitral decision on Transylvania. Two days ahead of the Vienna Diktat Hitler told Ciano, in von Ribbentrop's presence, that Hungary "had not earned anything through her own efforts but owed her revisionist victories solely to Fascism and National Socialism."[2] In response to the objections of the Romanian foreign minister, von Ribbentrop stated that Romania "was faced with the alternative of either losing not only all of Transylvania but in

---

[1] *Documents on German Foreign Policy 1918-1945*, Series D (1937-1945), vol. X, *The War Years. June 23—August 31,1940*. London, 1957, doc. 171, p. 218.

[2] *Ibidem*, doc. 407, p. 568.

addition head for a political catastrophe, or else agreeing to a solution which was reasonable although it involved relatively heavy sacrifices."[3]

The Vienna Diktat forced Romania to cede to Hungary a territory of 43,492 square kilometers which included the towns of Salonta, Oradea, Satu Mare, Baia Mare, Sighet, Dej, Năsăud, Bistrița, Cluj, Tîrgu Mureș, Miercurea Ciuc, Sfîntu Gheorghe, and others and a population of 2,667,007 inhabitants of whom 50.2% were Romanians, 37.1% Magyars, and the rest members of other nationalities.[4] Thus, the Romanians constituted the absolute majority of the population of the territory taken by Hungary. Moreover, in the eleven *județe* (counties) of that territory the Romanian population constituted a majority in eight and the Magyar in three.[5] Hitler himself had to concede, at a time when Hungary sought all of Transylvania, that "The structure of the population in the contested area — taking the most favorable view of Hungary's position — was as follows: as against $1\frac{1}{2}$ million Hungarians there were 3 million Rumanians and 600,000—700,000 Germans."[6]

In its angry reaction to the loss of Romanian territory, Romanian public opinion regarded the Vienna Diktat as a destructive action inspired by the Horthy regime. The historian Nicolae Iorga stated in the Crown Council that "This diktat represents the victory of a ruling Hungarian caste, it is the victory of the *grofs* who sought the Cluj region where they used to have large estates and big fortunes; we are dealing with a wealthy social class, and a shrewd one, which knew how to exploit a passing European contingency. For me matters are as follows: a social class belonging to the past and which has no future has been able to win by exploiting a moment

---

[3] *Ibidem*, doc. 408, p. 574.
[4] Arhiva M.A.E., București, Holding 71 Transilvania, vol. 42 (1940), f. 166.
[5] László Bányai, *Destin commun, traditions fraternelles*. Bucarest, 1972, p. 174. Budapest's figures were different, seeking to increase the size of the Magyar population and to decrease that of the Romanian. Nevertheless, the journal *Látohatár*, October 1942, page 220, shows the Romanian population at 48% and the Magyar population at 42%. See also Daniel Csatári, *Dans la tourmente. Les relations hungaro-roumaines de 1940 à 1945*. Budapest, 1974, p. 39, note 1.
[6] *Documents on German Foreign Policy*, vol. X, doc. 407, p. 569.

which is subject to all contingencies of the future, it has wanted to strip us of part of our country." [7]

Indignation and opposition to the Diktat were manifest throughout the country. All political parties and forces in Romania, including the Union of Magyar Workers in Romania *(Magyar Dolgozók Szövetsége* or *Madosz)* denounced the Diktat, organized mass demonstrations against it, and demanded military action in defense of Romanian territorial integrity and sovereignty. On the very day of the signing of the Diktat, on 30 August 1940, mass demonstrations took place in Cluj and similar demonstrations occurred also elsewhere in Transylvania. At Brașov, for instance, the masses stopped the automobile of the German consulate and trampled Hitler's portrait. Most newspapers and periodicals in Romania devoted continuous and extensive coverage to the protest of the Romanian people against the Diktat.

The validity of the annexation of northwestern Transylvania by Hungary was recognized exclusively by the fascist states and was categorically rejected by the anti-Hitlerite powers. On the evening of 30 August 1940 the London radio station stressed that that was a day of sorrow for the Romanian people, one of the darkest in the history of that nation. A monstrous deformation of Romania's borders was carried out in Vienna by the Diktat of the Axis powers.[8] Winston Churchill told the House of Commons on 5 September 1940 that the British Government did not intend to recognize territorial changes made during the war.[9] Secretary of State Cordell Hull, appearing before the Foreign Affairs Committee of the House of Representatives of the United States of America, first included the "splitting up of Rumania and the German occupation of the remaining portion of that country" among the acts of aggression committed by the fascist powers against European countries and then stated that "these seizures have been accomplished through a combined use of armed force applied from without and of an almost unbelievable amount

---

[7] Arhiva M.A.E., Holding 71 Transilvania, vol. 42, ff. 280—281.

[8] Arhivele Statului, București (State Archives, Bucharest), Holding of the Ministry of National Propaganda, Bulletins, Folder 34/1940, p. 205—206.

[9] Arhiva M.A.E., Holding 71 Transilvania, vol. 44, f. 265.

of subversive activity from within. Each of the invaded and occupied countries have been subjected to a reign of terror and despotism."[10]

The results of the partition of Transylvania were disastrous not only from a demographic standpoint but also from an economic one. Industrial establishments, the transportation network, deposits of raw materials, all were arbitrarily divided among the two parts of Transylvania with resultant chaotic conditions in the province's economic life. Hungarian-Romanian relations worsened. "Toward mid-October" — stated the Magyar historian Daniel Csatári — "it became evident that the Vienna Arbitration was not leading to a diminution of antagonisms but, rather, to an aggravation of tensions."[11]

Even before the conclusion of the formalities related to the taking over of the annexed territory the Horthy regime initiated, in northwestern Transylvania, chauvinist and anti-semitic policies and instituted a reign of military terror directed against Magyar communists and anti-fascists as well as against the Romanian population.[12] Many fascist and irredentist organizations, with headquarters in Budapest, sent emissaries to northern Transylvania. Inspired by chauvinist nationalism, such fascist and military outfits as *Rongyos Gárda* (Ragged Guard), *Nemzetörség* (National Guard), *Tüzharcosok* (Firing-line Fighters), or *Levente* (Paramilitary) engaged in terrorist activities, destructive actions, and abuse against the inhabitants under the very eyes and protection of Horthy's authorities. They had been exposed to the chauvinism and spirit of revenge contained in books such as that of the Hungarian writer Dücso Csaba, *Nincs kegyelem, Attila, Almos, Arpád ivadéka fel az uj hinfoglalásra* (Without Pity: Descendants of Attila, Almos, Arpad Forward to New Settlements), which was published in Budapest in 1939 with the approval of the authorities. The book called, *inter alia*, for the extermination of the Romanians by any means such as assassination, poisoning, arson as may be needed to attain this goal. In the book a young

---

[10] *New York Times*, 16 January 1941, p. 8.

[11] Csatári, *Dans la tourmente*, p. 43.

[12] Ladislau Bányai, *Pe făgașul tradițiilor frățești* (On the Path of Fraternal Traditions). București, 1971, p. 230.

man voiced his creed: "I will do away with any Wallach I shall meet. I will kill everyone of them. There shall be no mercy. I shall set fire to Wallach villages at night. I will put the entire population to the sword. I will poison the wells and will kill even babies in cribs." The conclusion to this program was: "I will do away with every Wallach and then only one nation will be in Transylvania, the Magyar, my nation, my blood! I shall render harmless future Horeas and Cloşcas. There shall be no mercy!"[13]

Numerous acts of violence were committed against the Romanian population by Horthy's administration. Hundreds of inhabitants of villages such as Ip, Trăsnea, Mureşenii Bârgăului, Şimleul Silvaniei were assassinated, over 200,000 Romanians were expelled, the Romanian Orthodox bishoprics of Oradea and Sighet were abolished, Romanian schools were closed, all sales and purchases of land and buildings which occurred between 1919 and 1940 were voided, the publication of Romanian newspapers and journals was prohibited pending the securing of new authorizations to publish, political leaders, cultural and religious personalities, as well as anti-fascists of Magyar, German, and other nationalities, were interned in forced labor camps or assigned to forced labor units.

Between 1940 and 1944 a policy of economic exploitation and of oppression of the working people, regardless of nationality, was practiced on that part of Transylvania which was annexed by Hungary. Moreover, the "final solution" of the Jewish question was put into practice with the resultant "destruction of nearly 150,000 Jews in northwestern Transylvania."[14]

[13] Dücso Csaba, *Nincs kegyelem, Attila, Almos, Arpád ivadéka fel az uj hinfoglalásra* (Without Pity: Descendants of Attila, Almos, Arpad Forward to New Settlements). Budapest, 1939, pp. 155—156.

[14] See "Remember", *Revista Cultului Mozaic*, No. 415, 1 June 1978. The anti-Semitic persecutions in northern Transylvania are described in great detail by the American historian Nicholas M. Nagy-Talavera — himself an eye witness — in *The Green Shirts & the Others*. Stanford, 1970. The author reveals that of the 825,000 Jews who lived in Hungary (converted to Christianity or members of the Jewish faith) only some 255 000 survived. Approximately 570,000 perished during the last few months of the war. On persecutions by Horthyists in northern Transylvania consult also the above-cited works by Daniel Csatári and Ladislau Banyai.

XVIII

THE LIBERATION OF NORTHWESTERN TRANSYLVANIA

In the summer of 1944 at a time when, as acknowledged by leaders of the allied powers, the end of the war was not yet in sight because of continuing stiff resistance by Hitler's armies, the national antifascist insurrection of 23 August occurred in Romania and Romania joined the war against Germany and her allies. Romania's decision was not an easy one to take as it could have entailed fatal risks. Germany still had an army of 612,508 men, or 26 divisions, on Romanian soil. For three days on end after the insurrection the German airforce bombarded Bucharest inflicting heavy losses while the German armies were moving toward the Romanian capital.

Romania's joining the allied forces contributed greatly to the final victory over the fascists. During the night of 23—24 August Romania committed all its armed forces to war against Hitler's armies. After eight days of heavy fighting, Romanian lands south of the Carpathians were freed; until 25 October Romanian forces, together with Soviet ones, continued the war of liberation of northwestern Romania. On 25 October 1944 the special communiqué of the Supreme Command of the Romanian armed forces, announced that in northwestern Transylvania the Romanian troops, cooperating closely with Soviet ones, have occupied the towns of Carei and Satu Mare and after that crossed the Romanian-Magyar frontier thus freeing all of Transylvania. [1]

The Romanian armed forces of some 540,000 men were fully engaged in the war against Hitler after 23 August 1944.

---

[1] *Pentru eliberarea Patriei. Documente* . . . *(23 august—25 octombrie 1944)* (Liberating the Fatherland. Documents . . .), București, 1972, p. 695.

The eighteen Romanian divisions fought for 260 days, during which time they penetrated the enemy lines as far as the vicinity of Prague and captured some 103,000 prisoners while losing some 170,000 men, killed, wounded, or missing. The Romanian armed forces prevented the Germans from carrying out plans for setting up defensive positions along fortified lines which would have allowed their gaining valuable time. The Romanians, by engaging in military actions both south and north of the Carpathians facilitated and allowed the speeding up of the Soviet offensive toward the Balkans and Central Europe. The Romanians' military activities contributed greatly to the destruction of Hungary's armed forces, which remained loyal to Hitler's Germany until the very end of the war, and to the freeing of Czechoslovak territories.

The population of northwestern Transylvania supported the Romanian military actions against Hitler's and Horthy's forces by sabotaging the supply of cereals and cattle, by slowing down rail and road transports, by opposing the evacuation of enterprises, by disrupting the communications system, by deserting forced labor units.

The freeing of Transylvania voided, de facto, the Vienna Diktat since the territories taken in 1940 were reintegrated. The Hungarian radio station "Kossuth," which broadcasted since 1941 from the Soviet Union, declared on 7 September 1944 that "The Vienna award regarding Transylvania had no legal power. On the contrary, whoever received territory as a gift from Hitler can count on losing it sooner or later. We have asked Hungary to declare the Vienna award invalid but we might as well have talked to the walls! Those who are destroying our country have not budged from their position that the Vienna award and Hitler are a source of law."[2]

On 12 September 1944 the Romanian government signed the Armistice Convention with the members of the anti-Hitler coalition represented by the U.S.S.R., Great Britain, and the United States. The Armistice Convention expressly recorded the voiding of the

---

[2] Gyula Kállai, *A magyar függetlenségi mozgalom 1936 — 1946* (The Magyar Movement for Independence) apud *Magazin istoric*, XIV, no. 8, 1980, p. 42.

Vienna Diktat. On 20 January 1945 the Italian government, headed by A. Bonomi, unanimously agreed, on the basis of the proposal submitted by the foreign minister Alcide de Gasperi, to declare the Vienna Diktat imposed on Romania on 30 August 1940 null and void. On 13 March 1945 a solemn session of the Romanian government, presided over by Dr. Petru Groza, was held in Cluj at which the instauration of Romanian governance in northern Transylvania was proclaimed.

In the period immediately following the end of the war representatives of the Magyar Popular Union in Romania — established through the reorganization of *Madosz* — stated repeatedly that they regarded the maintenance of the territorial order based on the frontiers set at the end of World War I as the one which assures best the free development and peaceful coexistence of all nationalities of Transylvania. Moreover, they also expressed their confidence in and satisfaction with the nationality policies of the democratic Romanian government headed by Dr. Petru Groza. Thus, the resolution of the central executive committee of the Magyar Popular Union, which met at Tîrgu Mureş, between 15 and 18 November 1945, declared, inter alia: "We are grateful to the Groza government for its efforts to guarantee our political rights, for the opportunities it provides for our being taught in our native language, and for the safeguarding of our economic interests." [3]

On 7 May 1946, the foreign ministers of the Soviet Union, the United States, Great Britain, and France in Paris declared the Vienna award of 30 August 1940 null and void. The border between Romania and Hungary, such as it was on 1 January 1938, was reestablished. That decision became article 2 of the Paris Peace Treaty with Romania in 1947. That decision, which restored northern Transylvania to Romania allowed the normalization of relations and the establishment of ties of friendship between Romania and Hungary.

Following the overthrowing of the dictatorship of General Ion Antonescu, most of the provisions of the Constitution of 1923 were reestablished on 31 August 1944. That constitution granted freedoms

---

[3] *Világosság*, Cluj, Nov. 21, 1945.

to all Romanian citizens regardless of ethnic origin or language. On 6 February 1945, Law No. 86 — known as the "Statute of Nationalities" — was adopted. It provided, in article 1, that "all Romanian citizens, regardless of race, nationality, language, and religion, are equal before the law and enjoy the same civil and political rights." The law not only contained general principles but also established measures and guarantees for the attainment of equal rights for all citizens. Moreover, the provisions which guaranteed equal rights to all citizens were elaborated and formalized by the constitutional decree No. 2218 and by Law No. 560 of 13 July 1946 with respect to electoral rights.

Following the reestablishment of diplomatic relations between Romania and Hungary, which had been broken on 30 August 1944, on 10 December 1946, a delegation of the Hungarian government headed by the prime minister, Lajos Dinnyés, visited Romania between 23 and 25 November 1947. That visit was, in fact, the first visit ever made by a Hungarian prime minister to Romania. At the end of the discussions, the prime minister emphasized that democratic Hungary does not believe in and does not tolerate revisionism of any kind. "The litigious problem between the two countries is regarded as solved. Magyar public opinion is grateful to the Romanian government for the national citizenship rights granted to the Magyars of Transylvania. Dr. Petru Groza's government has inaugurated a new era in Romanian-Hungarian relations."[4]

---

[4] *Scînteia* București, 25 Nov. 1947.

XIX

THE STATUS OF NATIONALITIES IN ROMANIA

Socialist Romania works on the assumption that it is necessary for nations and states to establish new relations based on total equality of rights, on mutual respect of independence, on every people's right to develop according to its will. Moreover, Romanian domestic and foreign policy is based on historic realities, on the fact that over the centuries close relations of collaboration have been established between Romanians and their compatriots of different nationalities, all of whom have worked together under normal as well as under difficult circumstances. The overall economic and cultural development recorded in territories inhabited by Romanians, Magyars, Germans, and other nationalities represents the result of common work and common struggles.

Romania has assured full equality of rights to all citizens of the country, regardless of their nationality, in the economic field, in labor, in the management of enterprises, in running all social affairs. The fundamental law of the Romanian state confirms, in legal terms the results of the entire social, economic, and political evolution which took place in Romania after World War II, a period of profound transformations for the country. Romania is a unitary national state in which people of other nationalities live and work beside the overwhelmingly Romanian population in a spirit of fraternity and unity. Thus, in 1980 Romania's population numbered 22,300,000 inhabitants of whom 88.14% were Romanians, 7.9% Magyars, 1.6% Germans, and 2.36% belonged to other nationalities.

The Constitution of the Socialist Republic of Romania of 1 August 1965 states, in article 17, that "the citizens of the Socialist Republic of Romania, regardless of nationality, race, sex, or religion, possess equal rights in all aspects of the economic, political, legal, social, and cultural life. The state guarantees the equality of rights of all citizens. No restriction of these rights and no discrimination in the exercising thereof because of nationality, race, sex, or religion are allowed. Any manifestation designed to set up such restrictions, chauvinist nationalist propaganda, incitement to racial or national hatred, are punished by law." Article 22 of the Constitution states that "in the Socialist Republic of Romania the coinhabiting nationalities are assured the unhindered utilization of the mother tongue as well as books, newspapers, journals, theaters, education at all levels in their own language. In districts also inhabited by a population of non-Romanian nationality, all organs and institutions use orally and in writing also the language of the respective nationalities and will appoint functionaries from these nationalities or from the ranks of citizens who are familiar with the language and way of life of the local population." Nationalist and chauvinist propaganda, incitement to race or national hatred, in accordance with article 317 of the Penal Code, are punished by law. The above-mentioned constitutional principles permeate the entire Romanian legislative system. They are not recognized only de jure. The Romanian state insists on application of these principles in the administrative sphere of all institutions, in all areas of activity ranging from political life and economic development to culture, education, and religious cults.

Citizens belonging to coinhabiting nationalities are full-fledged members of all political and social organizations and entities in Romania. In positions of leadership the numerical ratio of citizens to nationality is strictly observed. The Grand National Assembly of Romania consists of 369 deputies — elected on 9 March 1980 — of whom 332 are Romanian, 29 Magyar, six German and of other nationalities, in accordance with the country's demographic configuration. The same proportions hold true of members of other civic and political organizations. Similarly, in counties and localities inhabited by people of Szekler, Magyar, German, or other nationality the political and administrative organs reflect in their composition the national structure of the inhabitants. Presidents of the people's councils of counties, municipalities, towns, or of communes are elected from among members of these nationalities. Similarly, from the same ranks are appointed managers of industrial, agricultural, and commercial enterprises, directors of research institutes, school principals, heads of agricultural cooperatives, leading cadres of mass organizations, and so forth.

The coinhabiting nationalities of Romania are represented in the Front of Socialist Democracy and Unity through councils of working people organized according to nationality. The Council of the Working People of Magyar Nationality of the Socialist Republic of Romania coordinates the activities of fifteen county councils of that nationality. The work of the councils of coinhabiting nationalities is a function of the policies of the Romanian state seeking proper organizational forms for consulting with all working people, regardless of nationality, on important problems of domestic and foreign policy, for assuring the participation of these citizens in state and mass activities, in solving their own specific problems, in leading alongside their Romanian compatriots the efforts for the country's multilateral development.

In the view of the Romanian lawmaker the solving of the nationalities problems is expressly related to the general progress of the forces of production, to the elevation of the country's general level of civilization. Only by solving these fundamental problems may a framework be found also for solving the specific problems of one or another coinhabiting nationalities. Therefore, the balanced distribution of productive forces throughout Romania creates the conditions required for assuring true equality of rights and opportunities in all fields for all the citizens of the country. That, in the last analysis, is the prerequisite for achieving complete equality of rights for all citizens, irrespective of their nationality.

An essential role in achieving the socio-economic transformation for contemporary Romania has been played by the constant concern for assuring the development of all parts of the country through a judicious distribution of the forces of production. The administrative reorganization of 1968 created the bases for balanced countrywide economic development, for the industrialization of all counties of the country. With assistance from the state, the less developed Transylvanian counties made considerable progress. New industrial centers were established everywhere, including regions coinhabited by Romanians, Magyars, Germans, and other nationalities. Thus, in Covasna county industries reflective of technological progress, such as the electrotechnical industry, the machine building industry, the chemical industry, the metal processing industry, and the textile industry, were established. Similarly, in Harghita county new industrial estates were set up in Miercurea Ciuc, including a plant for caterpillar tractors, one for knitwear and spinning, an enterprise for toolmaking and spare parts manufacturing, a chipboard and furniture factory, and mining enterprises. An enterprise for cast-iron dies and mouldings, a furniture factory, a cotton spinning mill, and other establishments are to be found in Odorheiul Secuiesc. In Mureş county one may find major economic units in the energy field and in those of the chemical industry, metallurgical industry, wood-processing industry, light industry, foodstuff industry, and others. Between 1968 and 1979 overall industrial production increased in Covasna county from 1,626,000,000 lei to 7,627,000,000 lei; in Harghita county from 3,072,000,000 lei to 10,693,000,000 lei; in Mureş county from 10,478,000,000 lei to 26,990,000,000 lei. The industrial production in 1979 in Covasna, Harghita, and Mureş counties was, respectively, 43, 38, and 44 times that of the year 1938.

Moreover, it should be noted that significant increases occurred in the population of these administrative units. Thus, Covasna county had 152,563 inhabitants in 1930 as against 211,011 inhabitants in 1979.The respective figures for Harghita county are 250,194 and 341,230 and for Mureş county 425,721 and 607,116.

The establishment of new industrial units and corollary improved working conditions resultant from the modernization of production have contributed to the economic and cultural development of these counties. Between 1966 and 1979 21,574 dwellings were built in Covasna county, 30,248 in Harghita county, and 52,961 in Mureș county, either out of funds provided by the state or with financial assistance by the state.

Between 1976 and 1980 a large number of measures were adopted which led to an increase in the real average wages of all working people by better than 32%, and in 1978 the progressive reduction of the work week from 48 hours to 44 hours was initiated. Measures for raising the income of the peasantry, for improving old-age pensions and family allowances have also been adopted.

1. Painted vessel from the Neolithic Age (Ariușd, Covasna county).

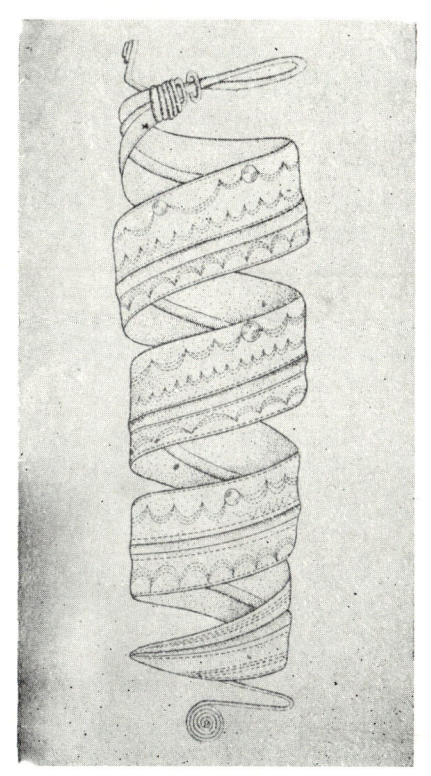

2. Spiral bracelet, bronze, 13th cent. B.C. (Pecica, Arad county).

3. Dacia in the time of Burebista.

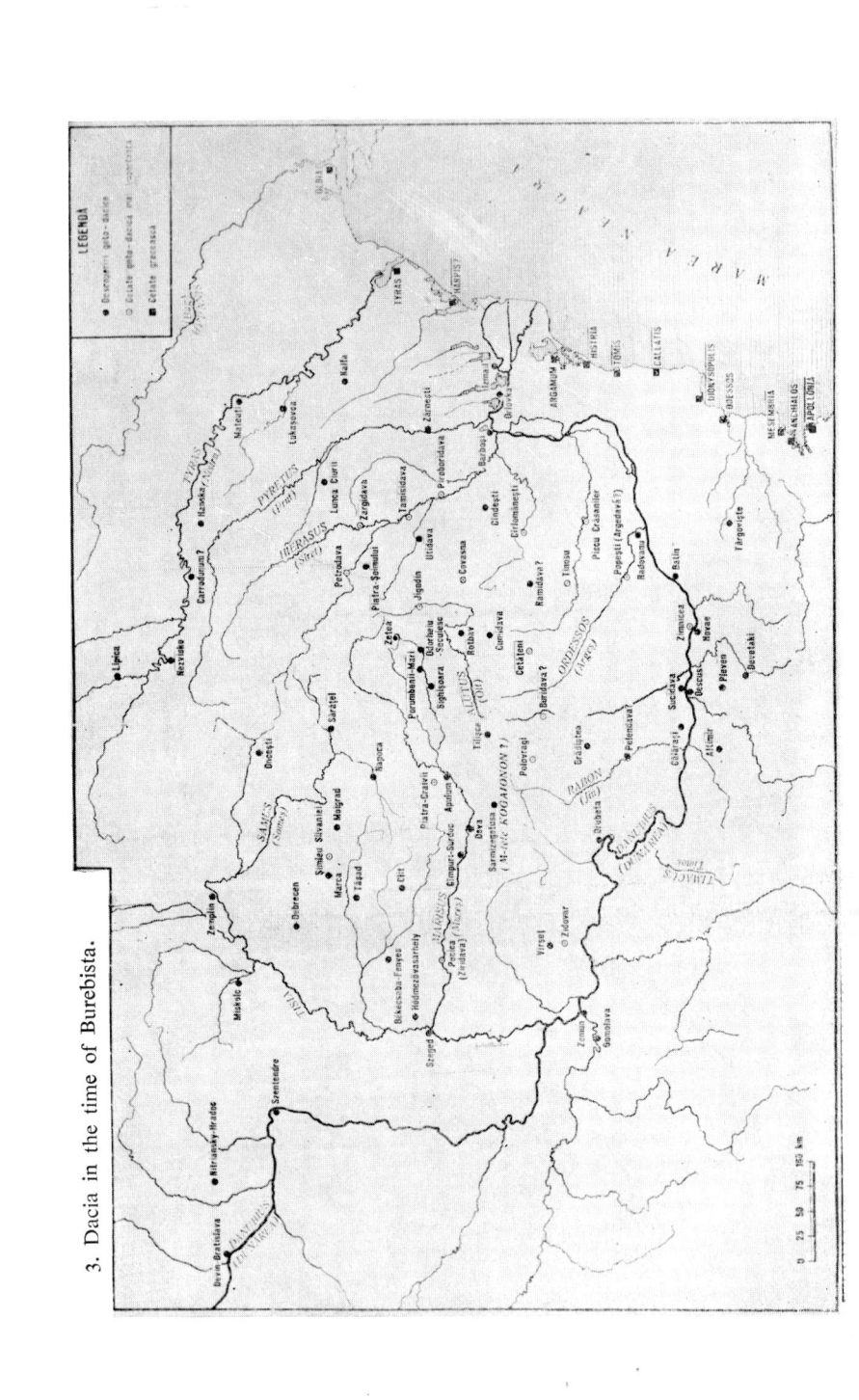

4. Dacian peasant *(comatus)*.

5. Dacian nobleman *(pileatus)*.

6. Fight between Dacians and Romans.

7. Dacians' return after the war. (Reliefs on Trajan's Column)

8. Roman Dacia.

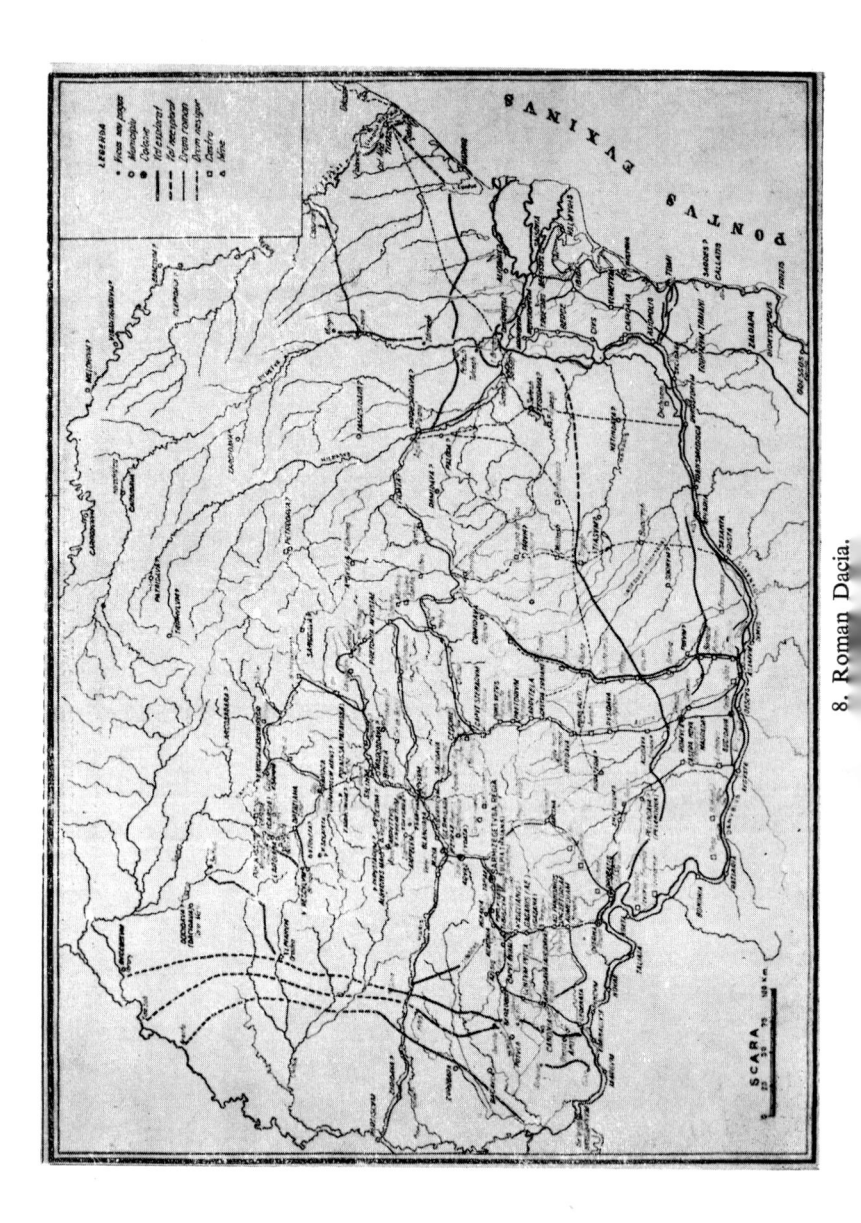

9. The migration period on Romania's territory.

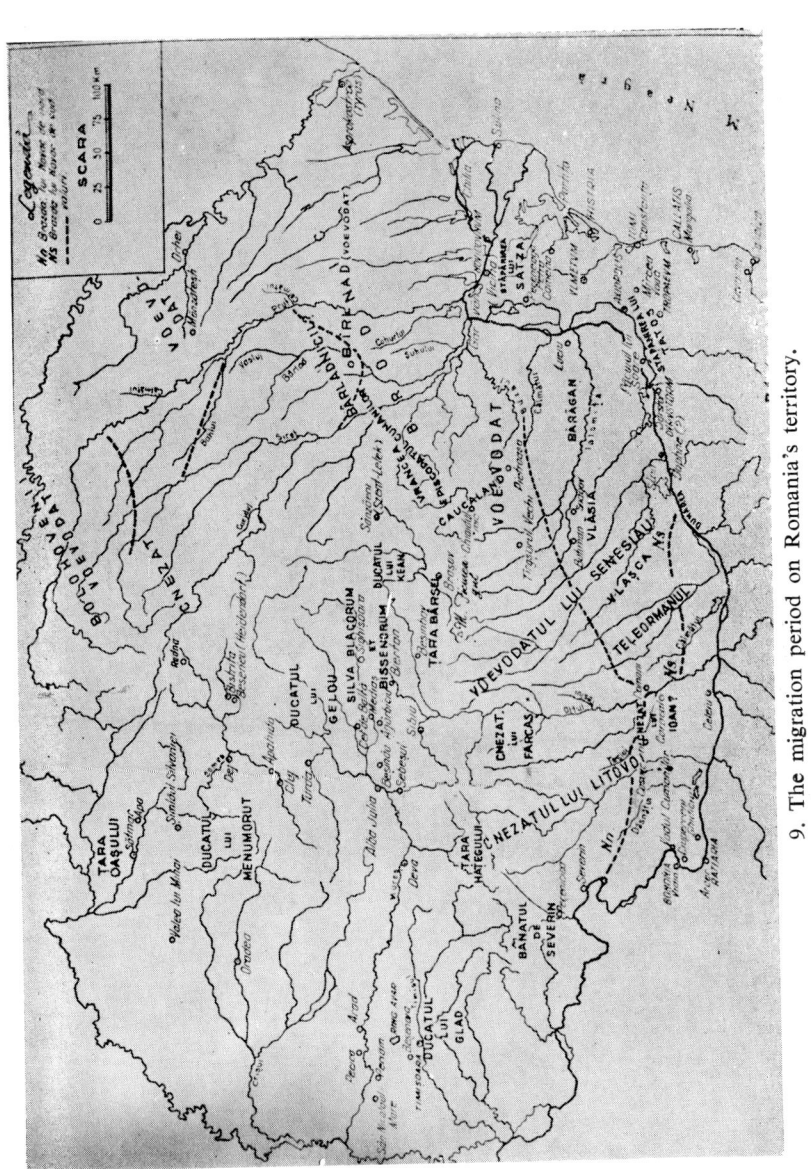

10. Page from Anonymus, *Gesta Hungarorum*, regarding the battles between Romanians and Hungarians in Transylvania.

De Terra Ult[ra]siluana

Et dum ibi diuti[us] morarent[ur], tunc Tuhutum pat[er] Horca, sic[ut] erat uir astut[us], du[m] cepisset audire ab incolis bonitatem terre Ult[ra]siluane, u[b]i Gelou q[ui]dam Blac[us] [domi]nium tenebat, cepit ad hoc hanelare [1], q[uo]d si posse e[ss]et, p[er] gr[ati]am ducis Arpad d[omi]ni sui t[er]ram Ult[ra]siluanam sibi et suis posteris acq[u]ireret. Q[uo]d et sic factum fuit postea. Nam, terra[m] Ultrasiluanam posteritas Tuhutum usq[ue] ad temp[us] s[an]c[t]i regis Steph[an]i habuerunt, et diucius habuissent, si minor Gyla c[um] duob[us] filiis suis Biuia et Bucna chr[ist]iani esse uoluissent et semp[er] [con]trarie s[an]c[t]o regi n[on] fecissent, ut in sequentib[us] dicetur. De prudentia Tuhuti.

Predict[us] u[er]o Tuhutu[m] uir prudentissim[us] misit q[ue]ndam uiru[m] astutum, patrem Opaforcos Ogmand, ut furtiue ambulans p[re]uideret sibi qualitatem et fertilitatem terre Ult[ra]siluane, et quales essent habitatores ei[us]. Q[uo]d si posse esset, bellum cu[m] eis co[m]mitteret. Nam uolebat Tuhutum p[er] se nom[en] s[ib]i et terra[m] aq[u]irere. Vt dicunt n[ost]ri ioculatores: om[ne]s loca s[ib]i aq[u]irebant, et nom[en] bonum accipiebant. Q[ui]d plura? Du[m] pater Ogmand, speculator Tuhutum, p[er] circuitum more uulpino bonitatem et fertilitatem t[er]re et habitatores ei[us] inspexisset, quantum human[us] uisus ualet, vltra q[ua]m dici po[te]st dilexit, et celerrimo cursu ad d[omi]n[u]m suum reuersus est. Q[ui] cum uenisset, d[omi]no suo de bonitate illius t[er]re multa dixit. Q[uo]d t[er]ra illa irrigaret[ur] optimis fluuiis, q[uo]r[um] no[m]i[n]a et utilitates seriatim dixit. Et q[uo]d in arenis eo[rum] aurum colligerent, et aurum terre illius optimv[m] esset. Et ut ibi foderet[ur] sal et salgenia, et habitatores terre illi[us] uiliores homines esse[nt] toci[us] mundi. Q[u]ia esse[n]t Blasij et Sclaui, q[u]ia alia arma n[on] haberent, n[is]i arcum et sagittas, et dux eo[rum] Geleou min[us] esset tenax et n[on] haberet circa se bonos milites, et auderent stare [con]tra audaciam Hungaro[rum], q[u]ia a Cumanis et Piicenatis multas iniurias paterent[ur].

[1] Corect: anhelare

11. Modern transcription of the same page.

12. Christian votive inscription of the 4th cent. A. D. at Biertan (Sibiu county).

13. The battle of Posada (1330) between Charles Robert of Anjou and Prince Basarab of Wallachia.

14

15

Two old Romanian monuments in Transylvania: Densuș and Strei (13th cent.).

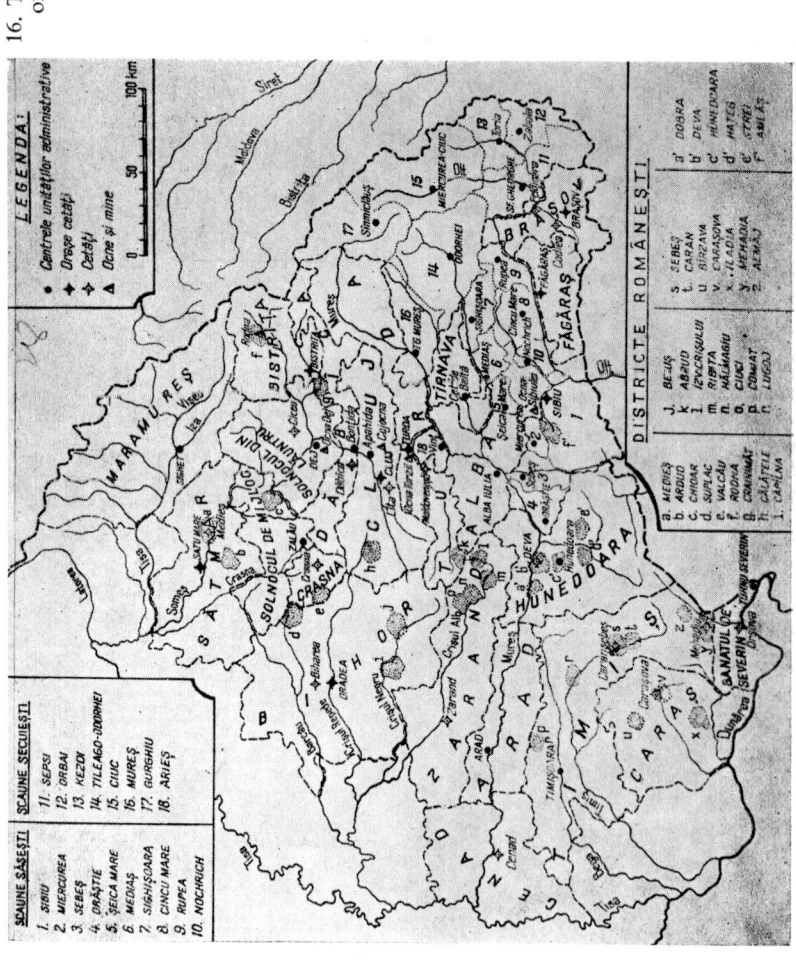

16. The political-administrative organization of Transylvania in the 14th century.

17

18

Nicolaus Olahus and Iohannes Honterus,
Transylvanian humanists.

19. John of Hunedoara, governor of Hungary and prince of Transylvania (1448).

20. Prince Michael the Brave (1593—1601), the first to accomplish the political union of the Romanians.

21. Bran castle (14th cent.).

22. The Black Church, Brașov (14th—15th cent.).

Horea, Cloşca and Crişan, the leaders of the uprising of the Transylvanian Romanians (1784).

24

25

26. Bishop Inochentie Micu, promoter of the political struggle of Transylvanian Romanians.

27. Title page of "Supplex Libellus Valachorum" (Cluj Edition).

28

29

Gheorghe Șincai and Petru Maior, representative historians of the Transylvanian School.

30. Simeon Bărnuțiu, Romanian thinker and revolutionary of Transylvania.

31. Nicolae Bălcescu, leader of the Romanian Revolution of 1848.

32. The Great Blaj Assembly of 3/15 May 1848.

33. The champions of the Romanian nationality of Transylvania (1848—1849).

34. Metropolitan Andrei Şaguna, outstanding figure in the political life of Transylvania.

35. Stephan Ludwig Roth, Saxon pastor, a friend of the Romanian people.

Ludovicu Mocsáry.

Desbaterile urmate dilele trecute în camer'a deputatilor Ungariei asupra introducerii limbei unguresci că studiu obligatoru în tóte scólele din tiéra au produsu unele incidinte memorabile. Unâ din aceste e, că între deputatii magiari s'a gasitu unul, durere numai unul, carele capabilu a se înaltiă deasupra netolerantiei sioviniştice a connationalilor sei, a avutu si curagiul a combate proiectul de lege, pentru cuventul că acela confiscă drepturile garantate si prin lege ale nationalitătilor nemagiare, si astfel în locu de a produce vr'unu bine, va nasce numai ura.

Acesta mare patriotu si adeveratu barbatu de statu, precum cetitorii nóstre au vedutu din unul trecutu al foii nóstre, e dl Ludovicu Mocsáry (Mociari).

Cu ce mare bucuria vor fi aflatu toti Românii acestu adeveratu evenimentu în parlamentul Ungariei, cu aceeasi placere grabimu si noi se aducemu tributul stimei nóstre facia de acestu barbatu, publicându-i portretul pe pagin'a acést'a.

Portretul unui unguru in „Familia"! De cinci-spre-diece ani de cându se publică fói'a acést'a, acuma se vede antéia-óra. Dar tocmai pentru că se ivesce pentru prima-óra, bucuri'a nostra devine si mai mare, amu puté-o numi epocala.

Vreti si schite biografice? Se spunemu dara, că acestu barbatu s'a nascutu la Kurtány, comitatul Nograd, în 1826. E proprietaru de paméntu.

A servitu la comitatu, înaintându până la postul de vice-comite, apoi s'a alesu deputatu, si aci începe activitatea sa interesanta pentru noi.

Ludovicu Mocsáry, prin positiunea sa sociala, prin educatiunea sa si latele cunoscintie în multifarielle ramificatiuni ale vietiei publice, prin sublimitatea ideilor si prin ardorea cu care sustine libertatile constitutionale, prin marturisirea si propagarea principielor adeverat democratice, cu unu cuventu, prin calitatile sale de omu de statu a devenitu si este unul dintre corifeii vietiei nóstre parlamentarie. Partid'a independenta l'a onoratu cu încrederea sa, alegându-l mereu de presidinte al clubului seu. Ceea ce îl distinge mai multu si îl înaltia în ochii nostri, este ari'a convictiunilor sale si nobilitatea sintemintelor, căci feritu de prejudetie, de esclusivismul si egoismul national, cu înaltul seu patriotismu îmbraciséza pe tóte nationalitatile patriei, si care după marele patriotu Francescu Deák si Paul Nyári, — înse mai resolutu decâtu acestia, — cu admirabila barbatia înfruntă si combate siovinismul connationalilor sei. Figur'a sublima a lui Ludovicu Mocsáry se înaltia dintre connationalii sei ca abietele muntilor pe de asupra piticilor junepeni.

Activitatea jurnalistica înainte si dupa reactivarea constitutiunii patriei, dar mai alesu cea parlamentaria a lui Mocsáry este destul de cunoscuta inteligintiei române, credu că si brosura lui scrisa în cestiunea de nationalitate este inca în viu'a memoria a Românilor.

Mocsáry apartinea mai nainte partitei centrului stângu, — al careia capu era actualul primu presidente al consiliului de ministri, dar elu, neputêndu suferi siovinismul acestui omu, s'a ruptu de acésta partita si s'a alaturatu la partid'a independintilor, — credêndu că aici principiele sale democratice vor gasi mai justa apretiuire, precum a si

Ludovicu Mocsáry.

36. Ludovic Mocsáry, Magyar politician, supporter of the nationalities rights.

37. The Romanian Memorandists (1894).

38. Romanian politicians who conducted negotiations with Count Tisza (1910).

39. The National Assembly of Alba Iulia (1 December 1918).

40. The Coronation Church of Alba Iulia.

41

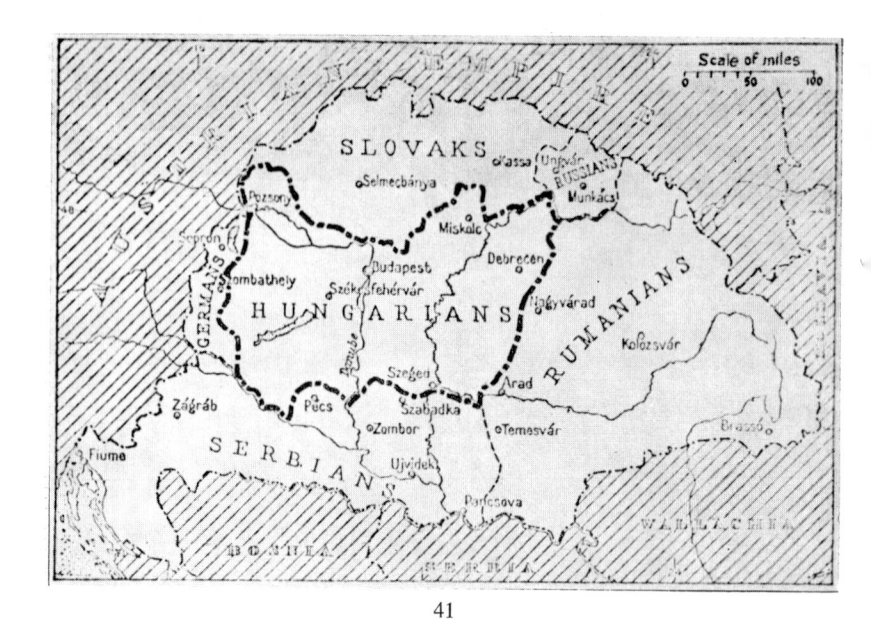

I. HUNGARY ACCORDING TO THE FEARS OF KOSSUTH IN 1850.
The territory of pre-war Hungary is unshaded. The heavy broken line indicates what Kossuth foresaw would be left of Hungary if non-Magyar elements were given autonomy and allowed to gravitate towards their co-nationals beyond the frontier.

42

II. HUNGARY ACCORDING TO THE TREATY OF TRIANON, 1919.
The former territory of Hungary is unshaded. The heavy broken line shows the present frontier of Hungary, which strongly resembles the frontier foreseen by Kossuth seventy years earlier.

43. Poet Octavian Goga.

44. Writer Liviu Rebreanu.

45. Victor Babeș, prominent personality of Romanian medicine.

46. Engineer Traian Vuia, pioneer of aviation.

47. The Romanian Lycée of Blaj.

48. The Polytechnical Institute of Timişoara.

49. The University House of Cluj-Napoca.

50. The interior of the National Theater of Timişoara.

51. The monument of Păuliş (Arad county) erected to the memory of the Romanian warriors fallen for Transylvania in 1944.

52. The iron-and-steel mills of Reșița.

53. The National Theater of Tîrgu-Mureș.

XX

CULTURAL POLICY AND THE COINHABITING NATIONALITIES

The cultural life of coinhabiting nationalities is an organic part of Romania's culture and enjoys material support from the state, providing a large network of cultural, scientific, and educational institutions in their mother tongue.

In the educational field, the equality of all citizens regardless of their nationality is evident in Romania because (1) every citizen is entitled to admission to any and all educational institutions and at all levels in accordance with his desires and aptitudes in relation to the country's requirements for economic and social development; (2) the use of the mother language of coinhabiting nationalities is assured in all educational forms or levels, parents having the right to freely choose for their children the school which provides instruction in the preferred language; (3) equal opportunities are provided to all citizens regardless of the language in which they were educated. Naturally, the educational institutions for the youth belonging to coinhabiting nationalities provide not only for the acquisition of knowledge in their mother tongue but also for acquisition of a correct knowledge of the Romanian language.

The coinhabiting nationalities have access today to nearly 3,300 educational establishments. In the school year 1979—1980 the school network with Magyar instruction for day students included 2,475 units of which 1,079 were for pre-school children, 1,276 general schools, and 120 lycées. In addition, in the sphere of general education there were 179 sections of extra-mural courses and 57 sections of evening lycées which provided opportunities for further study in Magyar for adults. Among lycées with instruction in Magyar there are some with long-standing traditions such as the lycées "Bethlen

Gabor" of Aiud, "Bolyai Farkaș" of Tîrgu Mureș, "Brassai Samuel" of Cluj-Napoca, and others. In higher and university education many of the disciplines are taught in Magyar: e.g. the "Babeș-Bolyai" University, the Music Conservatory "Gheorghe Dima," and the Institute of Fine Arts "Ion Andreescu," all in Cluj-Napoca as well as at the Institute of Medicine and Pharmacology and the Drama Institute "Szentgyörgyi Istvan" in Tîrgu Mureș.

During the school year 1979—1980, 264,810 children and pupils attended day schools of whom 54,161 were in pre-schools, 180,173 in general schools, and 30,476 in lycées. In addition 7,768 pupils attended sections of extra-mural courses and 5,940 evening sections. Thus, the total number of young people of Magyar nationality pursuing their education in the Magyar language was 278,518.

During the same school year, 9,622 Magyar students were enrolled in institutions of higher learning. Of these 5,433 were enrolled in higher technological education, 452 in agricultural studies, 600 in economic studies, 1,414 in medical studies, 114 in legal studies, 1,392 in pedagogical studies, and 217 in art-oriented fields.

Several scientific and technological fields offered by institutions of higher learning are taught only in Romanian. However, university courses designed to train young Transylvanian Magyars for work related to the preservation and development of Magyar culture, language, history, literature, and art — and even courses in medicine offered at the Institute of Medicine and Pharmacology in Tîrgu Mureș—are given in the mother tongue. It is also worth noting that every year hundreds of young people who have pursued most of their studies in the Magyar language graduate from the "Babeș-Bolyai" University of Cluj-Napoca.

During the school year 1979—1980, the faculty of schools with instruction in Magyar amounted to 13,016 of whom 2,264 were pre-school teachers, 4,220 school-teachers, 5,119 teachers in gymnasiums, 1,413 high school teachers and vocational instructors, while in universities the faculty providing instruction in Magyar numbered 740, including 81 professors, 111 lecturers, 266 readers, and 262 assistants. One hundred textbooks for use in schools with Magyar as the language of instruction were published in 1980 in 1,200,000 copies with a value of 10,000,000 lei.

The situation is the same with respect to education in the German language which is provided through kindergardens, general schools, lycées, and departments or sections in establishments of higher learning.

A significant contribution to vocational training and improvement is made by cultural-scientific universities, by technological-economic lectureships, and by circles for applied technology functioning in enterprises, clubs, and houses of culture of trade unions, whose numbers have increased markedly also among the coinhabiting nationalities. Of the 17,903 courses offered by the 2,503 cultural-scientific universities which functioned during the school year 1978—1979, for instance, 862 with 33,768 attendants were given in Magyar and 116 with 3,814 attendants in German.

In the field of publishing, eleven firms issue literary works of all genres and provide books in the languages of coinhabiting nationalities through the printing of over 300 titles in approximately 3,000,000 copies. During the period 1970—1979 the "Kriterion" publishing house alone published 1,055 titles in Magyar, 360 in German, 118 in Serbo-Croatian, 80 in Ukrainian, 24 in Yiddish, and two in Slovak. In addition, 600—700 titles in 500,000—700,000 copies are books in Magyar and German imported annually from abroad.

Literary works of genuine value are written in the languages of coinhabiting nationalities. The scope, circulation, and financial support of Magyar and German literature produced in Romania are unsurpassed in history.

Important achievements are recorded also in literary criticism and history in which fields numerous studies, monographs, and works of synthesis have been published. Through the activities of scholarly researchers works of leaders in scientific thought have been rediscovered and circulated and the writings of the most representative authors, poets, and men of culture of the interwar years as well as anthologies of progressive journals of that period, have been reedited. By becoming accessible again during the last three decades, the cultural achievements of the coinhabiting nationalities have facilitated the development of scientific awareness of the history and character of the literary creation of these nationalities and have contributed to the flourishing of their cultural life. Special attention has been paid to literature for children and youth.

An important role in the guidance and discovery of young writers belongs to literary clubs and circles such as the literary club "Gaál Gábor," of Cluj-Napoca, the literary club of the journal *Igaz Szó*, of Tîrgu Mureș, the literary clubs "Ady Endre" and "Adam Müller-Guttenbrunn," of Timișoara, or the literary circle sponsored by the newspaper *Karpaten Rundschau*, of Brașov. The discovery and promotion of young talent is energetically undertaken by the literary journals *Igaz Szó, Utunk, Neue Literatur* as well as by the "Forrás" collection devoted to that purpose and published by the "Kriterion" publishing house. Among those who receive prizes from the Writers' Union one finds, every year, the best writers in various literary genres, young and old, and translators from among the ranks of coinhabiting nationalities. In 1979 the great prize of the Writers' Union was awarded to a writer in Magyar, Méliusz Jozsef. During the years 1976—1979 alone, 18 other writers in Magyar, 9 in German, 4 in Serbian, and one in Yiddish received prizes from the Writers' Union, not to mention prizes awarded by county writers' associations, by the Academy of the Socialist Republic of Romania, by youth organizations, and so forth.

Scientists, researchers, technological and university cadres from the ranks of coinhabiting nationalities contribute to the development of science and technology in Romania. They hold leading positions in academies of science, in institutes, wherever they are needed for research in the fundamental and applied sciences. Much research is done by historians of Magyar and German nationalities which has resulted in important works on subjects which contribute to mutual understanding of the problems, traditions and concerns of the Romanian and coinhabiting nationalities. The activities of scholars result in publications on the philosophy and history of culture, on popular art and technology, on folklore which appear as books, articles, and studies.

Prestigious works by linguists have appeared also, such as "Magyar Orthographic Dictionary," "Romanian-Magyar Dictionary," "Historical Dictionary of the Magyar Vocabulary in Transylvania," by

Szabó T. Attila, "Dictionary of Saxon Dialects in Transylvania," by Bernhard Capesius, and others.

The political, socio-cultural, and literary-artistic press and the radio and television broadcasts in Magyar, German, and Serbian play an important role in informing and educating the public. Fifty-four newspapers and periodicals are published in the languages of co-inhabiting nationalities—thirty-four in Magyar, eight in German, three in Serbian, one in Ukrainian, one in Armenian, and seven in 2—3 languages. In 1978 the total number of issues exceeded 125,000,000. This publication network includes socio-political dailies, such as *Elöre* and *Neuer Weg* — organs of the Front of Socialist Democracy and Unity; 14 county newspapers; socio-cultural journals, such as *A Hét, Korunk, Müvelödés, Volk und Kultur;* literary-artistic publications, such as *Igaz Szó, Utunk, Neue Literatur, Knijevni Jivot;* journals for children and youth, such as *Napsugár Jóbarát* and the weekly supplement of the paper *Neuer Weg, Raketenpost;* departmental publications, such as *Munkásélet, Tanügyi Ujság, Méhészet Romániában;* journals of ethnographic and linguistic studies and research, such as *Nyelv- és irodalomtudományi Közlemények, Forschungen zur Volks- und Landeskunde;* university juornals, and others. There are also five religious publications of the Catholic, Protestant, and Mosaic cults. Supplements and topical yearbooks are also published for the readers' edification and education by journals; they include the quarterly supplement *Tett* and the yearbook "Fatherland, birthplace, nationality" issued by the journal *A Hét*, the yearbooks of *Korunk* and *Utunk,* and the calendars published by the dailies *Elöre* and *Neuer Weg.*

The central station of the Romanian Television has weekly programs in Magyar and German. In addition to news, these programs include debates and polls on topics related to ethics, social problems, technological and scientific problems, specialization of cadres; there are also programs for children and youths as well as operatic, theatrical, and concert performances. A substantial part of the programming is devoted to problems related to cultural-artistic activities, whether amateur or professional, to the presentation of traditions and folklore specific to coinhabiting nationalities, to promotion of national languages, to new books and other publications, and so forth. Radio stations, likewise, broadcast from Bucharest, Cluj-Napoca, and Tîrgu Mureș some 40 hours per week in Magyar, from Bucharest and Timișoara 13 hours in German, and from Timișoara also 6 hours in Serbian.

The spiritual life of coinhabiting nationalities is reflected also in the performing arts. There are now nine theaters, the Magyar opera of Cluj-Napoca — established in 1948 — and four marionette theaters which offer performances with the aid of state subventions. These establishments have well trained performers whose number is increased every year by young actors graduating from the Drama Institute "Szentgyörgyi István," of Tîrgu Mureș, and by directors trained at the Theater and Cinematography Institute "I. L. Caragiale," of Bucharest. New cadres for the two German-language theaters of Timișoara and Sibiu are assured through the existence of a German program for actors at the "I. L. Caragiale" Institute and by the sending of scholarship holders to the advanced school for directors in Berlin. The repertory of Magyar-language theaters includes, in the first place, classical Magyar works but also Romanian and universal dramatic masterpieces. The achievements of institutions of the performing arts are rewarded by the state. Thus, in 1978, for instance, the Magyar and the German state theaters of Timișoara were awarded the order "Cultural Merit," first class, on the occasion of their twenty-fifth anniversary.

The works of painters, sculptors, and other artists from the ranks of coinhabiting nationalities are displayed both in personal shows and in general regional, national, or international exhibitions. Magyar and German personalities in political history, science, and culture, such as Apor Péter, Apáczai Csere János, Johannes Honterus, Petöffi Sándor, Kriza János, Bolyai János, Hermann Oberth, Adolf Menschendörfer, Nagy István, and others, have been commemorated in Romania as well as by the UNESCO.

The Constitution of the Socialist Republic of Romania provides, in article 30, for the freedom of conscience. Everyone is allowed to have or not to have a religious faith; religious organizations function without hindrance, the freedom to practice one's religion is guaranteed. The activity of the 14 religious cults in Romania is reflected by the large number of places of worship. The state provides the means for the upkeep and repairing of these places. It should be noted that during the last few years alone the faithful of Magyar nationality built 16 churches and 7 houses of worship and modernized 591 churches and 17 houses of worship.

In Romania there are six theological institutes with 1,590 students and 13 theological seminaries with 2,082 students. Of these, three theological institutes, with 314 students, and two theological seminaries, with 118 students, belong to coinhabiting nationalities. The cults publish 19 periodicals of which 5 are printed in the languages of coinhabiting nationalities. These are *Reformatus Szemle* and *Keresztény Magvető*, in Magyar, *Kirchliche Blätter*, in German, "Bulletin of the Orthodox Serbian Vicarage," in Serbian, "The Journal of the Mosaic Cult," published in Romanian, Yiddish, and Hebrew.

The freedom of religion is also assured by the fact that the cults possess houses, buildings, agricultural lands, printing shops, factories and shops for producing religious objects; they also manage denominational cemeteries. The cults have their own budgets which consist of donations by the faithful, sales of religious objects, and state subventions. The cults' personnel receive salaries — equivalent to those of members of the educational system — which are paid in part from state subventions and in part by the respective churches. The cults entertain relations with various foreign churches and participate in international religious activities.

Architectural monuments, works of art, old libraries, all related to the cultural and historic past of coinhabiting nationalities are organized, restored, and given importance with substantial contributions by the state. These include the Sf. Mihail church of Cluj-Napoca, the episcopal palace of Oradea, the Roman-Catholic cathedral of Alba Iulia, the historic libraries "Samuel Teleki" of Tîrgu Mureş, that of the Protestant college of Cluj-Napoca, that of the Bethlen family of Aiud. An important monograph on the Magyar libraries of Transylvania by the Hungarian scholar Jakó Zsigmond of Cluj-Napoca has been recently published in Magyar as well as in Romanian by the "Kriterion" publishing house. The publication is illustrative of the interest displayed by Romanian culture in the works of Magyar culture in Romania.

The Romanian government is committed to ensuring the participation of all citizens, regardless of nationality, in the shaping of the country's present and future. As stated by the President of the Socialist Republic of Romania, Nicolae Ceaușescu: "Today we can proudly say that among our political achievements we can include the just resolution of the nationalities problem and, as a consequence thereof, the strengthening of the unity of the entire people in the struggle for socialism, for the building of the multilaterally-developed socialist society in Romania."

The contemporary history of Romania bears witness to that truth.

APPENDIX A

1785

Jacques-Pierre Brissot

Seconde lettre d'un défenseur du peuple à l'Empereur Joseph II, principalement sur la révolte des Valaques: où l'on discute à fond le droit de révolte du peuple

> Those who wrote against
> the unfortunate Wallach people [1]

Tous ceux qui ont écrit sur la révolte des Valaques semblent avoir conspiré contre le malheureux peuple, pour t'encourager, Prince à punir les chefs par d'horribles supplices, à serrer les liens du peuple J'en ai vu même qui plaisantoient sur les roues où le *demagogue** Horiah (comme ils l'appeloient) devoit finir ses jours. Loin de moi la doctrine abominable de ces monstres qui prostituent leurs plumes pour le malheur des peuples. Puisse un jour le ciel accumuler sur leurs têtes toutes les horreurs de la servitude qu'ils prêchent avec tant de scélératesse!

«Le respect qu'on doit à l'humanité est sacré comme celui qu'on doit aux dieux, & quiconque se joue dans ses discours, des droits

---

[1] Throughout the appendices, headlines belong to the authors.

* *Démagogue* est le nom que les aristocrates, les despotes ou plutôt leurs mercenaires écrivains donnent aux chefs du peuple qui veulent le tirer d'oppression. Il est très plaisant de voir ces lâches valets du despotisme donner de l'encens au démagogue Washington, parcequ'il est vainqueur de la tirannie, & envoyer à l'échafaut tel autre démagogue ou défenseur du peuple parcequ'il a succombé. Si Horiah eût réussi, ils auroient mis sur sa tête une couronne de lauriers.

que les hommes tiennent de la nature, doit expier ce sacrilège pour le mépris des sages & le poignard des opprimés»*.

On the present status of the Wallaehs, and on
the reasons which led them to revolt

Mais pour juger des vrais caractères de cette révolte, il faudroit connoitre une foule de faits que le public ignore. Il faudroit connoitre l'état actuel des Valaques, leurs mœurs, leur gouvernement, les impôts qu'ils paient, la proportion de ces impôts, avec le produit de leur culture. Il faudroit connoitre les causes qui les ont portés à se révolter, les injustices qu'on leur a faites, le degré d'oppression sous lequel ils gémissent. Il faudroit connoitre l'origine, la marche de cette révolte, son histoire; il faudroit savoir s'ils sont réellement coupables de tous les meurtres qu'on leur prête, s'ils ont été forcés pour leur sûreté ou leur vengeance à verser tant de sang; si en le répandant ils ont épargné celui des êtres foibles, des femmes & des enfans. Voilà ce que n'ont point dit les gazettes qui racontent avec une exactitude si ennuieuse tant de petits faits étrangers à l'histoire.

Et quand elles en parleroient, devroient-elles être crues? Ces gazettes ne sont-elles pas des sources corrompues? Ne sont-elles pas dans chaque gouvernement vendues à ses chefs, à ses Ministres? Leurs rédacteurs soudoïés, n'ont-ils pas ordre de taire les vérités favorables aux peuples, comme les fautes de l'administration? N'est-ce pas enfin une nouvelle conspiration pour tromper la postérité sur les événemens du siècle présent?

Excepté les Etats unis de l'Amérique & l'Angleterre, partout ailleurs, le peuple n'a point, comme ses chefs, des gazetiers à ses ordres ou qui défendent ses intérêts. C'est partout le *triste pecus* qu'on tond, qu'on égorge, & les gazetiers sont comme ces tambours qui, placés

---

* Telephe, liv. **7.** Je cite ici ce Roman pour rendre honneur à un écrivain peu connu, qui mérita plus de l'être, & qui a vigoureusement défendu les droits de l'humanité; je regrette seulement que pour les venger, il ait pris le masque de la fable, il est tems que les apôtres de la liberté se montrent hardiment, M. Plumejcat mort depuis quelques mois, tenoit un rang distingué parmi eux.

autour du taureau brûlant de Phalaris, remplissoient l'air de leurs cris bruyans pour étouffer les cris des victimes.

> No confidence whatsoever
> in newspaper accounts of the revolt

On ne doit donc ajouter aucune foi aux récits au moins suspects des gazetiers dans l'insurrection des Valaques. On aura exagéré leurs fautes, tu leur misère, travesti leur résistance en attaque, leur défense en assassinats; on aura tout calculé, tout bien combiné pour les rendre criminels, & pour justifier ces illustres *Nobles* qui les tyrannisoient & l'administration qui se prêtoit à leurs vexations.

Oh! si j'avois en main les pièces de ce procès, si j'avois observé pendant quelque tems le sort de ce peuple, si je l'avois suivi dans ses travaux comme dans ses tourmens, si j'avois vécu dans ses chaumières, dans le sein de la terre souvent son unique asile, si j'avois assisté à la conspiration, & j'avois été témoin de ses supplications au souverain pour terminer sa misère, de ses combats pour la finir lui même, avec quelle vérité, quelle énergie je le défendrois! Mais loin du théâtre de cette guerre, n'aïant sur elles que les détails ou arides ou mensongers des gazettes, seul, presque sans lumière, & n'étant soutenu que par le zèle qui m'entraîne à défendre le peuple partout où je le vois se débattant sous le couteau de l'oppression, que puis-je faire, sinon de prouver par le récit même du parti qui a triomphé, que les Valaques étoient fondés dans leur insurrection? Oui, je ne veux consulter ici que le récit d'un de ces écrivains qui s'est plu à couvrir de boue Horiah & ses partisans. C'est ce récit à la main que le prétens prouver l'injustice de leur condamnation.

> The Wallachs were right in their revolt

Les Valaques se sont révoltés; donc ils avoient raison de se révolter. Voilà ma première prèuve; elle paroitra singulière, elle n'est que naturelle.

Qu'on se rappelle ce que j'ai dit ci-devant que le peuple malgré l'oppression sous laquelle il étoit courbé se laissoit difficilement entrainer à la révolte, qu'accoutumé à un certain ordre de choses, il n'en sortoit qu'avec peine; quelque détestable qu'il fût, qu'il s'habituoit à croire que sa place marquée par la tirannie étoit réellement marquée par la divinité. Il se laisse d'autant moins encore entrainer à la révolte qu'il est plus ignorant. Car alors il est loin d'imaginer qu'il est l'égal de ces nobles qui le tirannisent, du Souverain qui domine ses tirans; il est loin d'imaginer que la nature lui donne le droit le résister à l'autorité, que la divinité même lui ordonne. Il croit ses fers divins & il les baise, aïant à peine la force de les soulever.

Quand donc un pareil peuple entrant en fureur, brise ses chaines malgré ses prejugés civils & religieux, il en faut conclure qu'il a eu raison de se révolter, puisqu'il s'est révolté.

Le récit du gazetier confirme cette conclusion à l'égard des Valaques: L'état de barbarie & d'ignorance que je viens de peindre est précisément leur tableau.

Soumis depuis une foule de siècles à un despotisme plus ou mois sévère, suivant le caractère des Princes dont il portoit le joug, ce peuple a langui constamment dans la misère & l'ignorance. Le régime féodal, dont les traits horribles sont effacés presque par toute la terre, conserve encore toutes ses rigueurs au sein de cette misérable contrée. On y retrouve ces vieux Barons anglois, ces Comtes françois qui, cantonnés dans leurs petits forts, regardoient leurs hommes de glèbe comme des meubles dont ils disposoient à volonté, dont ils pouvoient jouer, vendre, aliéner la liberté, les sueurs & la vie même.

The violence and oppression of the nobility

Suivant le récit, *l'aversion d'Horiah pour les gentilshommes, & le désir de se délivrer & ses compatriotes des violences de la noblesse, ont été les premiers ressorts de son entreprise.*

On convient dans un autre endroit que ces nobles avoient *forcé leurs sujets à la révolte par leur oppression.*

Ces deux phrases suffisent, ce me semble, pour justifier la révolte des Valaques. Ils étoient malheureux, voilà le titre qui les armoit, & si ce titre ne vaut rien, Brutus étoit coupable en chassant Tarquin de Rome, l'autre Brutus en perçant Cesar, & tous ceux qui ont délivré leur patrie du joug des Néron, des Phalaris, & des monstres qui s'abreuvoient dans le sang des hommes, tous ces héros, dis-je,

Appendices

méritoient l'opprobre & la mort, au lieu des lauriers dont tous les siècles les ont couverts.

Je ne répéterai point tout ce que j'ai dit ci-devant sur le droit de révolte. Je me borne à suivre ici la narration du gazetier, à en extraire les faits principaux. L'application des principes sera facile à leur faire.

Horea at the Emperor

Il paroit, suivant cet écrivain, qu'*Horiah mit les griefs de sa nation sous les yeux du Monarque, qui lui promit de mettre fin à la tirannie.*

Ce procédé n'est point d'un barbare; il est celui d'un bon sujet qui, avant d'avoir recours aux droits que la nature lui donne, veut éviter une rupture, & l'effusion du sang. Averti par lui, Prince, tu ne pouvois pas ignorer qu'une partie de tes sujets languissoit dans l'esclavage, & si tu n'as pas rempli la promesse que tu lui avois faite, il faut croire que les difficultés se sont multipliées sous tes pas: il faut croire que la servitude ne pouvant finir que par la destruction du régime féodal, tu n'eus pu couper toutes les têtes de cette hidre; il faut croire que tu as été forcé de renfermer dans ton sein tes vues bienfaisantes.

Mais enfin quelqu'ait été le motif qui a pu t'empêcher de soulager les malheureux Valaques, au moins il faut avouer que leur conduite étoit régulière. Horiah avoit fait son devoir, c'étoit à toi à faire le tien. Tu ne l'as pu, l'oppression a toujours subsisté, c'est à dire un état de guerre qui dure depuis un tems immémorial, & alors les Valaques ont pu résister. Cet état de guerre les a replacés dans l'ordre naturel. Ils étoient tirannisés par des monstres, ils ont pu leur résister, & s'ils ont été coupables, c'est d'avoir attendu si longtems à venger la nature outragée.

The Emperor's visit in Wallach regions

Le voïage que ton désir de connoitre tes États t'avoit porté à faire, il y a quelques années, dans cette contrée, t'avoit mis à portée d'y connoitre la misère qui y régnoit. Je me rappele la substance d'une requête qui te fut présentée par un Valaque. Elle étoit ainsi:

«Prince, nous travaillons quatre jours de la semaine pour nos maitres; le cinquième est pour le Ministre du Seigneur, le sixième pour nous, & le septième nous l'emploïons à célébrer le dimanche».

Non, les nègres ne trainent pas dans nos isles une existence aussi infernale; il faut qu'un pareil peuple meure de désespoir, ou qu'il égorge ses tirans *.

Horea's aim

Je ne dirois rien du stratagème emploïé par Horiah pour engager ses compatriotes dans la conspiration qu'il tramoit. Il vouloit les délivrer de la servitude & ce but légitimoit tout.

On lui prête de grandes cruautés, en commençant la révolte. Encore une fois, loin du païs, je ne puis verifier ces faits. Cette opération est facile à tes juges, & pour décider si ces meurtres sont punissables, il faut s'astreindre aux régles que j'ai ci-devant posées. Etoient-ils nécessaires au projet? Etoit-ce représailles? Point de peine. Etoit-ce un jeu de barbares? La mort: toutefois si le supplice de la mort est encore dans tes Etats la peine de l'assassinat.

L'Angleterre & la France ont été, dans les siècles passés, témoins de révoltes de païsans qui, comme les Valaques, étoient las de la tirannie de leurs nobles. Le massacre qu'ils en firent, paroissoit mérité & fut plus considérable. Il ne fut point puni. Un peuple doux jusqu'alors ne peut être à la fois un amas d'assassins. Il se fait justice alors parce qu'il n'y en a point pour lui, & si on le punissoit, ce seroit d'être juste.

The position of the Government

Le gazetier dit que *le gouvernement provincial, embarrassé d'abord, chercha à apaiser les troubles en partie par la douceur, en partie par les menaces.*

Il ne falloit ni douceur, ni menace; il falloit la justice. En pareil cas, la douceur semble un piège & l'est souvent, & la menace aigrit.

Il envoia des Commissaires & l'Evêque grec aux rebelles.

---

* Tous les seigneurs valaques n'étoient pas de petits tirans. Il en est sans doute qu'il faut excepter. Les papiers publics en ont cité plusieurs qui au milieu de la révolte traitoient humainement leurs vaissaux.

Appendices

Des Commissaires sont des juges, & un peuple dans l'état de guerre ou de nature, ne connoit ni supérieur ni juge; il faut traiter à l'égal.

Un Évêque! toujours des prêtres sur la scène politique! probablement parcequ'ils connoissent parfaitement l'art de séduire & de tromper.

Si j'étois sur le trône & que j'eusse eu le malheur de mécontenter mon peuple, j'en agirois plus franchement avec lui. Je traiterois moi même avec les mécontens, moi, sans armée, sans commissaires, sans courtisans & surtout sans prêtres. Je leur dirois: mes amis, je ne veux point être votre bourreau, on me nomme votre père, je veux l'être. Montrez moi vos maux, je les guérirai. Je ne leur offrirois ni pardons, ni parchemins. C'est aux sujets malheureux à pardonner &, au lieu d'un parchemin, j'abolirois sur le champ les abus, les vexations, & je bannirois les petits tirans. Voilà comment en agit un père avec ses enfans, un grand homme avec des hommes.

Le gouvernement chercha à tirer d'erreur les rebelles.

De stratagème d'Horiah n'étoit que l'accessoire; mais il falloit prouver aux Valaques qu'ils étoient dans l'erreur en se croïant malheureux.

Il promit enfin un prix de trente florins pour la capture de chacun d'eux.

Fiez-vous donc à présent aux douceurs, aux caresses, à l'équité des gouvernemens! quand vous ne tombez pas dans leurs pièges, alors ils se démasquent; & quel autre moïen infâme emploient-ils? c'est d'invoquer la perfidie, la trahison de leurs sujets contre les rebelles. Ils encouragent les premiers au crime à prix d'argent.

Encore une fois, est-ce là la marche d'un gouvernement équitable? Ne doit-il pas offrir justice & se borner là?

Observons que le gouvernement, en mettant la tête des rebelles à prix, les forçoit à la guerre, aux meutres, à la perfidie, & légitimoit ainsi ces crimes dont la honte devoit retomber sur lui seul.

The nobles are guilty

Continuons le récit. «Les gentilshommes qui crurent que le gouvernement manquoit d'activité pour réprimer la rebellion, se crurent autorisés eux-mêmes à une insurrection. Ils allèrent en troupe contre les rebelles, les assomèrent où ils les rencontrèrent, & firent rouer, pendre, décapiter, empaler les prisonniers sans autre forme de procès. Par là les gentilshommes agirent directement contre les mesures du gouvernement, qui tendoient à ramener la paix par l'indulgence, & ils augmentèrent par leur propre violence la fureur des révoltés.»

Les conséquences de ces faits sont aisées à tirer.

J'ai donc eu raison de dire ci-devant que dans une insurrection le parti de l'oppression étoit toujours le plus féroce & le plus violent. Les rebelles donnent la mort, les tirans prolongent la vie des leurs prisonniers pour jouir de leur douleur. Les nobles faisoient rouer, empaler, lorsque les Valaques se bornoient à tuer. Certes, si les deux partis étoient coupables, il faut avouer que celui des nobles l'étoit doublement, & cependant on verra que les païsans furent punis, lorsque pas un des nobles ne le fut.

The revolt was legitimized by the tortures inflicted by the nobility

Il est évident encore que par les supplices qu'ils infligeoient, les nobles auroient légitimé la révolte, si d'ailleurs elle n'eût pas déjà été légitime.

Il est évident encore qu'en la supposant criminelle, les supplices infligés par ces nobles, & la mort de plus de 600 Valaques devoient l'avoir suffisamment expiée.

Il est évident enfin que les nobles agirent directement contre l'intention du gouvernement. Ils étoient donc coupables & cependant encore une fois ils ne furent pas punis.

N'oublions pas de remarquer cette *indulgence* prêtée au gouvernement. ... Indulgence déplacée. ... Un Souverain doit être juste d'abord. L'indulgence suppose des vices, des défauts à supporter, à excuser; & c'étoit le peuple qui avoit besoin d'indulgence pour le gouvernement.

The rulers' interests

«Le gouvernement considéra que les Valaques composent les deux tiers des habitans du païs, qu'ils ont entre eux une liaison étroite, que ce sont eux qui cultivent leur païs, & qu'en les exterminant, la province deviendroit un désert. Il apprécia mieux les hommes, & sentit la préférence qu'il faut donner à deux mains qui travaillent sur une bouche qui ne fait que consommer».

Quel horrible calcul, que celui devant lequel le gazetier paroit s'extasier? Ainsi donc ce n'est pas sur la bonté de la cause, mais sur l'utilité des parties par rapport au gouvernement, que celui-ci mesure la justice qu'il rend, la protection qu'il accorde! ainsi il auroit favorisé la tirannie des nobles, s'il avoit calculé que les nobles fussent plus utiles à son intérét? N'avois-je donc pas raison de dire dans ma précédente lettre, que l'intérét seul du gouvernement & non le bonheur des sujets le dirigeoit dans ses démarches? Je disois: ne nous laissons point éblouir par ce mot de prospérité de l'État, qu'on cite avec emphase, & voïons ce qu'il cache. Plus d'hommes, plus de soldats; plus d'hommes, plus de capitation; plus d'hommes, plus d'industrie & de taxes; plus d'hommes enfin, plus puissant est le Prince, plus riche est son trésor, & c'est de ce trésor que sort la foudre qui doit écraser le malheureux assez éclairé pour être pénétré de sa situation, &c.

«Animé de ces principes, le gouvernement a desapprouvé l'insurrection & a fait aux nobles des représentations».

Quoique le motif du gouvernement fût erroné, quoiqu'il n'eût pas le droit de se montrer indulgent ou moderé, quoiqu'il eût dû d'abord être juste, cependant il faut le louer de la modération qu'il montra, des représentations qu'il fit aux nobles. Autrefois, quand on traitoit avec des rebelles, on commençoit par élever des échafauts, où l'on amonceloit innocens & coupables sous la hache du bourreau. Telle étoit la justice d'alors. On n'est pas si cruel aujourd'hui. C'est toujours un gain sur la tirannie. Peut-être à la fin sera-t-on juste avec les rebelles, & ton âme éclairée, Prince, est assez grande pour n'en pas donner l'exemple à demi, comme tu l'as fait dans cette circonstance.

Car pourquoi le pardon offert? Pourquoi cette défense d'insurrection? Pourquoi cette autre défense d'infliger la peine de mort?

Revolt is a right of the oppressed

Le pardon, je l'ai déjà dit, s'accorde à un coupable, & les Valaques n'étoient que malheureux. L'insurrection est un droit pour l'opprimé

& on ne peut le lui interdire qu'en détruisant l'oppression. La peine de mort étoit une injustice & il ne falloit pas s'en abstenir par grâce, mais par justice.

Tu nommes des Commissaires, tu les envoies aux rebelles & en même tems tu fais assembler des régimens pour les envelopper, pour les ferrer.

N'étoit-ce donc pas une contradiction? N'étoit-ce pas détruire l'effet des propositions de paix faites aux rebelles? N'étoit-ce pas leur dire: rentrez dans votre esclavage, sinon le sabre vous y fera rentrer?

Les Valaques te crioient: nous sommes opprimés, nous mourons de faim, nous expirons sous le fouet de nos bourreaux: tes commissaires, tes régimens leur disoient: soïez opprimés, mourez de faim, expirez sous le fouet, sinon on vous sabrera, on empalera vos prisonniers & vos chefs. Grand Dieu! quelle justice, quelle douceur! quelle indulgence! & les gouvernemens paroissent tout surpris de ce qu'on ne croit pas à leurs promesses, de ce qu'on persiste dans la révolte? Ils s'en indignent, ils menacent de supplices, comme s'ils expioient leur oppression sourde par la violence ouverte; comme s'ils n'aggravoient pas ainsi leur propre crime.

The rulers' promises

Le gouvernement à la vérité promettoit de faire droit sur leurs griefs, de remédier aux abus, si l'on rentroit en paix.

Mais ne sait-on pas que dans tous les tems l'administration s'est jouée de pareilles promesses? L'orage passé, elle rit de la crédulité du peuple, & continue à l'écraser. Lisez l'histoire, vous trouverez mille exemples semblables où le peuple fut toujours dupe de sa bonne foi.

Je veux tirer de cette promesse du gouvernement un raisonnement sans réplique contre lui. S'il promettoit de remédier aux abus, il y en avoit donc, & les Valaques avoient eu raison de se révolter, & l'on avoit tort de les punir.

Appendices 109

The rebels' proposals

Sur les promesses faites par le gouvernement, les Valaques présentèrent les points de Capitulation suivans :
1°. Le comitat & tous les propriétaires nobles prêteront serment, sur la croix probablement, d'exécuter le projet proposé.
2°. Il n'y aura plus de noblesse.
3°. Qui peut avoir un emploi impérial en doit vivre. Les gentilshommes comme le petit peuple doivent païer la contribution.
4°. Les nobles doivent quitter leurs possessions.
5°. Les terres des nobles, conformément aux ordres que Sa Majesté donnera, seront réparties entre les païsans.

They are just

Y a t-il rien de plus raisonable, de plus naturel que toutes ces propositions? Par la seconde on tarissoit à jamais la source de l'oppression, en anéantissant la noblesse. Car il faut être convaincu qu'elle ne peut exister qu'avec la servitude & que pour le malheur des sociétés.

La politique la plus saine semble avoir dicté la troisième. Les valets de la Cour doivent être païés par la Cour &'non pas du sang des peuples. Les tirans dont les besoins surpassent les revenus, ont adopté cette manière d'acquitter les services qu'ils reçoivent. Ils accordent à leurs serviteurs l'exploitation de leurs sujets, comme on accorde la tonte d'un troupeau, ou la coupe d'une forêt. C'est ainsi que le grand Mogol & les Soubas indiens en usent; ils donnent des contrées entières à piller à leurs soldats.

Est-il quelqu'un de bon sens qui nie que tous les membres d'un Etat en doivent supporter le fardeau proportionellement à leurs facultés? N'est-ce pas une vérité demontrée, & si l'on s'en écarte encore dans plusieurs roïaumes, c'est que l'injustice qui a poussé de profondes racines est difficile à extirper; c'est que le peuple est trop traitable, dit Hutcheson.

La 4me & la 3me proposition sont celles qui auront sans doute le plus révolté, & cependant elles étoient encore dans la nature. Car chaque Valaque rentrant par la révolte dans l'état de nature, avoit droit à la terre que ses sueurs avoient arrosée & fertilisée.

Chacun avoit droit à une propriété & comme toutes les propriétés étoient entre les mains des nobles, il est évident qu'on ne pouvoit restituer à chaque rebelle ce qui lui appartenoit, qu'en dépouillant les nobles de ce qu'ils avoient usurpé.

Reviewing the history of the rebellions

Si ces propositions étoient injustes, il faut dire que les déclarations des États unis d'Amérique étoient aussi injustes, car elles sont exactement les mêmes. L'égalité parfaite que la Pensilvanie, par exemple, veut entre tous les membres, détruit toute idée de noblesse & entraine la conséquence que les impôts doivent être également répartis.

Si les Américains ont été fondés à se révolter, parcequ'on vouloit les taxer sans leur consentement, à plus forte raison avoient-ils ce droit, les Valaques, qui n'avoient ni propriété ni liberté, qui étoient à la merci de maitres impitoïables.

Sous le règne tirannique de Richard II en Angleterre, cent mille païsans se révoltèrent à cause des impôts excessifs dont on les accabloit. Ils marchèrent vers Londres aïant un Maréchal nommé Walter à leur tête. Il demanda au Roi que tous les esclaves fussent mis en liberté, que les communes fussent ouvertes aux pauvres comme aux riches. Walter fut tué, on accorda aux païsans une charte de liberté, & les chefs de la rébellion furent punis. La charte fut bientôt après révoquée.

En parcourant l'histoire des révoltes, je me suis convaincu de trois vérités: la première, que le peuple ne se révoltoit que quand il étoit opprimé, & qu'il ne demandoit jamais que des choses justes; la seconde que les Souverains promettoient avec facilité, tenoient avec difficulté & souvent ne tenoient point leurs promesses; la troisième enfin qu'en accordant au peuple ce qu'il demandoit, on punissoit presque toujours ses chefs, ce qui est une contradiction bien révoltante.

Tel fut le sort des Gracchus à Rome, de Walter, de Cade en Angleterre, des Facio, de Lemaître à Genève, de Horiah en Valachie.

People can always gain through uprisings

Il faut ajouter cependant que, quoique par la mauvaise foi des gouvernemens, les peuples rebelles ne jouissent pas longtems de

Appendices

ce qu'ils ont obtenu, ils gagnent toujours à la révolte. A chaque pas qu'ils font, ils détachent un anneau de leur chaîne & quand ils ne feroient qu'effraïer les chefs, que les rendre moins entreprenans, plus circonspects, ils auroient toujours gagné.

Qu'on compare les demandes modérées du peuple à celles de ses maîtres. Du sang & des fers, tout se réduit là.

Et cependant si quelqu'un avoit droit de demander du sang & des fers, n'étoit-ce pas le Valaque? Martirisé depuis tant de siècles, n'auroit-il pas été excusable de demander le sang de son maître barbare & de le mettre aux fers?...

Mais non... l'homme du peuple est généralement bon & juste & plus juste que les riches si fiers de leur or & de leurs lumières.

Le célèbre Vendôme disoit qu'en observant les querelles des muletiers avec leurs mulets, il avoit toujours remarqué que la raison étoit du côté de ces derniers. Voilà précisément l'histoire du peuple & des gouvernemens.

The Wallachs' proposals were not accepted

Je reprens la suite de l'histoire des Valaques. Leurs propositions ne furent point acceptés. On se résolut à la guerre: les Valaques se partagèrent la contrée qu'ils habitoient, mais ils ne jouirent pas longtems de ce partage. On envoïa une armée contre eux, on les relança dans leurs bois, dans leurs montagnes, & enfin, dit le gazetier, on reçut la nouvelle que le soulèvement avoit été apaisé, sans qu'on eût emploïé la force ou versé du sang... Horiah abandonné de ses compatriotes fuit, sa tête est mise à prix, il est bientôt découvert, dénoncé, livré, & il expire dans les supplices.

On a calculé que les rebelles avoient tué près de 300 hommes, que 5 à 600 d'entre eux avoient péri.

Que conclure de ce récit, sinon que la force a terminé cette révolte. Tous les hommes ne sont pas des Saguntius. Les Genevois représentants ont cédé leur liberté aux François & aux Savoïards; les Valaques l'abandonnent aux troupes de l'Empereur. Et cela doit être moins étonnant. La pusillanimité dont on contracte l'habitude dans l'esclavage, quitte difficilement une âme qu'elle a dégradée. Abandonné de l'espoir de vaincre, l'esclave aime à croire aux promesses de ses maitres, il aime à croire que cet essai de son pouvoir & de ses droits le ramenera vers la douceur.

Que pouvoient d'ailleurs les Valaques sans vivres, sans argent, sans armée réglée, sans chefs, sans intelligence, contre des troupes aguerries? On voit la mort certaine en combattant. On entrevoit encore quelques raïons de vie en cédant, & l'espoir de la vie la plus misérable a des attraits pour le vulgaire. Il faut avoir une âme extraordinaire pour lui préférer la mort, & dans ce siècle où chacun capitule avec ses besoins & sa liberté & les croit bien païés par quelques jouissances, ces âmes privilegiées sont rares. Pour faire trembler les tirans, on n'a qu'à mépriser la vie. Si les opprimés avoient du courage, il n'y auroit bientôt plus d'oppresseurs; mais par un nouveau malheur le courage ne germe que dans un cœur libre. La lâcheté repose avec la servitude.

Les princes ne l'ignorent pas, & voilà pourquoi ils ont si peu de scrupule à emploïer la force pour ramener sous leur joug le troupeau qui s'en écarte. S'ils s'attendoient à trouver des Brutus, des Cromwel, des Ludlow, ils seroient plus économes du sang & des pleurs de leurs sujets.

Le succès presque général qu'ils rencontrent dans leur système de violence, les y confirme & parvient à le légitimer à leurs yeux. On ne tarde pas à croire juste ce qui nous est utile.

Ton âme est trop éclairée, Prince, pour te laisser aveugler par les illusions du despotisme. Non, tu n'ignores pas que la force emploïée pour calmer le mécontentement est un moïen illégitime & qui répugne au droit de tes sujets, un moïen qui aggrave le délit du gouvernement.

The language of despair used
by over 600,000 Wallachs

La force ne rend point heureux ceux qui ne l'étoient pas; elle ne fait que doubler leur désespoir, en leur offrant comme éternels, comme irrémédiables, les maux qu'ils endurent. Ils doivent se dire: Je mourrai donc enchaîné sur ce lit de douleur que j'arrose de mon sang & de mes sueurs. Il n'est plus d'espoir pour moi. Le ciel

Appendices

semble combattre avec les hommes contre ma liberté. Cédons donc & souffrons.

Tel est le langage de désespoir que tiennent plus de six cens mille Valaques à l'heure où j'écris cette lettre. Grand Dieu! qui peut soutenir cette idée? Un seule homme faire le malheur de milions d'hommes & n'être pas accablé de remords!... Non, je ne sais quelle âme on revêtit en montant sur le trône ou en approchant ses degrés; mais il me semble à moi, que si j'avois le malheur d'y être, je ne goûterois pas un seul instant de repos avant d'avoir essuïé les larmes de ces malheureux.

Lorsque dans un soir d'été, un doux crépuscule fraichit l'atmosphere & la dore de brillantes couleurs, j'aime à contempler ce spectacle, à adorer mon maitre, je jouis & je le remercie de ces suaves jouissances que procurent une conscience pure & la méditation. Mes regards tombent sur une chaumière & je me dis: Là peut-être sont des époux qui ont travaillé tout le jour pour un morceau de pain que de petits enfants affamés se disputent. Là peut-être une mère verse des larmes, un père s'attriste de ne pouvoir élever sa famille... Ces idées me percent l'âme & je voudrois être la divinité pour rendre tous les mortels heureux.

Prince; qui davantage approche de la divinité que les Rois? Ils ont le pouvoir de faire des heureux & ils sont comptables à la divinité de toutes les larmes qui se versent dans leur empire & qu'ils pourroient essuïer.

The mistake of putting down
the rebellion by force

Tels sont sans doute tes sentimens. Et si tu as emploïé la force contre les Valaques, c'est pour te donner le tems de mûrir le grand projet qui doit les rendre à la liberté. L'exécution seule de ce projet peut expier la faute d'avoir étouffé le soulèvement par la violence.

Who will be able to repair the injustices
which have been committed?

Mais qui pourra réparer l'injustice faite aux chefs de ce soulèvement? Si les griefs de leurs compatriotes étoient fondés, si tu as promis toi-même de leur faire droit, de réparer les maux soufferts par les Valaques, si par cette espérance donnée, tu as calmé les rebelles; si loin de les regarder comme criminels, tu les regarde comme les victimes d'une oppression que tu veux abolir, comment les chefs de ces Valaques seroient-ils plus coupables? Comment méritoient-ils la mort? Ils combattoient pour la même cause. Ils devoient partager le sort de leurs Compatriotes: ou s'ils méritoient la mort, il ne devoit y avoir ni grâce ni justice pour la nation entière.

Les as-tu punis comme assassins? Les as-tu punis comme rebelles? Je ne puis décider la première question. Il me semble cependant que Horiah pouvoit verser sans crime le sang des tirans qui avoient tant de fois versé celui de ses Compatriotes, qu'il pouvoit user de représailles pour tant de meurtres commis dans cette guerre par les nobles eux-mêmes. Au moins si je condamnois Horiah comme assassin, je ferois monter sur l'échafaut les ennemis qui, comme lui, avoient trempé leurs mains dans le sang, & il ne me seroit pas difficile de prouver que ces derniers étoient bien plus criminels. Et puisqu'aucun n'a païé de sa tête ses atrocités, je suis fondé à conclure, sans entrer dans aucun détail de faits, qu'il étoit injuste de punir de mort Horiah comme assassin, lorsqu'on laissoit la vie aux nobles assassins.

The wrongness of the sentence

A-t-il été puni de mort comme rebelle? il est bien plus aisé de prouver l'injustice de sa sentence; & il suffit de se rappeler les principes sur le droit de révolte que j'ai posés ci-devant. Car [1°] les Valaques & leur chef Horiah avoient le droit & l'ont encore de se révolter, puisqu'ils sont privés des droits inhérens à l'homme, de ces droits imprescriptibles, inaliénables.

En second lieu, les Valaques étoient malheureux & gémissoient sous l'oppression, & ce second titre est presqu'aussi fort que le premier pour fonder l'insurrection.

3°. Ils avoient exposé plus d'une fois leurs griefs au gouvernement.

4°. Ils ne demandoient au gouvernement que des choses raisonables, que de rentrer dans leurs droits de propriété & de liberté. Jamais on n'avoit eu égard à leurs plaintes. On leur refusoit leurs demandes. Maltraités par le gouvernement, ils avoient donc le droit de renoncer à la société, ils rentroient dans l'état de nature, & ils avoient le droit de prendre les armes pour assurer leur nouvel état; ils avoient le droit de former une nouvelle société, de la soustraire à l'ancien gouvernement.

Arrêté, dit le code de Pensilvanie, *que tous les hommes ont un droit inhérent à leur nature de former un nouvel Etat dans des contrées vacantes, ou sur un territoire qu'ils auront acquis toutes les fois qu'ils pourront augmenter leur bonheur.*

<small>The Wallachs were right in their revolt</small>

Si cet arrêté est vrai, il l'est pour tous les peuples, & les Valaques étoient fondés dans leur révolte. S'il est faux, toutes les Monarchies de l'Europe doivent s'élever contre les États-unis d'Amérique, doivent proscrire leurs constitutions, doivent rejetter, abhorrer toute espèce d'alliance avec ces insurgens, comme avec des infâmes. Mais loin de proscrire cette alliance, elles la recherchent; loin de proscrire ces constitutions, elles les laissent imprimer & circuler dans leurs Etats & les sanctionnent par le consentement qu'elles donnent à cette publication. Donc les Valaques ont pu suivre des principes qu'ils ont vu généralement adoptés, qui l'étoient même par les Monarques, & qui sont d'ailleurs dictés par la nature.

Prince, tes semblables se jouoient autrefois de la vie de leurs sujets: on a vu des Monarques condamner de leur propre autorité des hommes, uniquement parcequ'ils leur déplaisoient. L'histoire a flétri le nom de ces Princes. Ils sont dans la classe ordinaire des tirans.

<small>Accounting for those who were the victims of force</small>

Tu n'as ici versé le sang que de deux hommes. Mais ils peuvent être innocens comme assassins; ils le sont comme rebelles. Le premier, je viens réclamer sur leurs cendres pour ces cendres mêmes. C'est mon devoir, & j'en ai le droit. Car puisque le peuple n'a plus de défenseur, c'est le devoir du philosophe de veiller sur les Magistrats du peuple, d'éclairer leurs jugemens, de chercher à connoitre si ceux qu'ils frappent méritent d'être punis ou d'être vengés. Le tien,

Prince, est d'écouter ma réclamation, de soumettre à un examen plus approfondi ce procès jugé probablement d'après les principes erronés de la vieille jurisprudence des Monarchies.

Le premier, je defens encore la cause des Valaques sacrifiés à la force dans cette insurrection, & je le répète, ils ont le droit de se révolter, tant qu'ils seront esclaves & malheureux. Les punir d'exercer ce droit, c'est les punir d'être hommes.

Ou tu es convaincu de cette vérité, ou tu n'y crois pas encore; dans le premier cas, tu ne dois plus perdre un seul moment pour rendre la liberté aux Valaques, pour leur accorder une propriété. Si tu crois devoir respecter l'usurpation de leurs nobles, si tu ne veux pas leur faire restituer, il est un moïen simple de faire justice à tous. Permets à ces Valaques d'émigrer dans tes autres Etats, ou ailleurs. Accorde-leur des terres, & ils quitteront en foule la terre de servitude. Et que deviendront alors ces gentilâtres dans leurs forts environnés de déserts, de friches couverts de ronces? Quand il faudroit louer des vaisseaux pour exporter le peuple en Amérique, tu le devrois, si tu n'avois pas d'autre moïen de le rendre au bonheur. ... Si ces nobles opposent la force à tes vues humaines, rends ce païs à lui-même; laisse aux esclaves le droit de s'armer contre leurs tirans & la querelle sera bientôt terminée.

For the rights of humanity and liberty

Si tu n'es pas convaincu de ces vérités, ton devoir est de lire ce plaidoïer pour la liberté. S'il te laisse des doutes, consulte tes jurisconsultes, tes politiques, mais ne te bornes pas à les consulter dans le secret; ils n'ont que trop souvent égaré les Monarques, quand ils n'avoient pas à craindre le grand jour. Qu'ils publient leur opinion, qu'il soit libre de la discuter, de se nommer, & j'offre d'entrer dans la lice, de résoudre leurs objections. Que dis-je, une foule de bons esprits paroitront aussitôt pour défendre les droits de l'humanité & de la liberté.

⟨Jacques-Pierre Brissot⟩, *Seconde lettre d'un défenseur du peuple à l'Empereur Joseph II, sur son Règlement concernant l'émigration, et principalement sur la révolte des Valaques; où l'on discute à fond le droit de révolte du peuple*. Dublin, MDCCLXXXV, pp. 65—89.

APPENDIX B

1791

Supplex Libellus Valachorum Transsilvaniae

Blessed August Emperor!

Considering that in governing this Empire Your Majesty's loftiest aim and most justified intention is that everywhere the rights both of man and of the civil society be extended first of all to all the members who form it through their union and carry out the task of maintaining it, at the cost of their lives and property and that one part of the citizens should not break loose ⟨from the other⟩, to deprive it ⟨the latter⟩ by force of its rights and oppress it; in this petition, the Romanian nation living in the Great Principality of Transylvania comes publicly before the throne of Your Majesty to beseech that their ancient rights be restored to them, those rights which are the natural concern of all the citizens, and of which they were despoiled arbitrarily in the last century, as shall be shown below, not by any law but by the iniquity of the times.

> The ancientry of the Romanian nation
> in Transylvania and her origin

The Romanian nation is by far the oldest of all the nations of the Transylvania of today, as it is common knowledge and has been proved by historical evidence and by a tradition never interrupted, by the resemblance between the languages, the customs and habits, that they descend from the Roman colonies repeatedly brought here in Dacia at the beginning of the second century by Emperor Trajan, with a very large number of veterans, to protect the Province.

The descendants of Emperor Trajan were the masters of Dacia for several centuries. Under their uninterrupted rule, the Christian faith, the rite of the Eastern Church, spread to these parts due to the endeavours of bishops Protogenes, Guadentius, Nicetas and Theotinus, especially in the 4th century, as shown by the whole history of the Church.

Her continuity

Already in the 3rd century, barbarian tribes began to endanger the life ⟨of the people⟩ in this rich Province of the Roman Empire and they succeeded for a time in settling in some parts of it. Still they were unable to go so far as to completely wipe out the name or power of the Romans. One thing is certain, in the 6th century several fortified towns situated especially on the banks of the Danube were under the domination of the emperors of the Eastern Roman Empire, while the inner parts of the Province were filled with such a great number of Roman inhabitants that, as early as the 7th century, they shook off the foreign yoke and founded a state of their own.

It was especially the part of Dacia today named Transylvania that had that good luck and after getting rid of the rule of other tribes, the Roman inhabitants were ruled by princes of their own nation, until the coming of the Hungarians.

There remained and has lasted to this day the vestige of the domination which certain Slav tribes, among other foreign peoples that passed after them in these parts, had on the Roman inhabitants of Dacia, ⟨namely⟩ the name of *Vlachs* or *Walachs* which for Slav peoples meant, according to the evidence supplied by Lucius the Dalmatian and Cromer the Pole, any *Roman, Italic,* or *Latin:* it was preserved in later times only for the Roman inhabitants of Dacia.

The arrival of the Magyars

When by the end of the 9th century the Hungarians led by ⟨their⟩ Duke Tuhutum invaded Transylvania, the Roman inhabitants of these ⟨parts⟩ had a changed name, *Vlachs,* according to the earliest writer of Hungary, Anonymus, the notary of King Bela; at their head there was their own Duke Gelu, who enjoyed supreme power

Appendices

but was unlucky in the war he started against the Hungarians for the defence of his own country, as he lost his throne and his life in that fight.

After the sad fate of ⟨their⟩ prince, the Roman inhabitants of the Province who are mentioned as *Vlachs* no longer resisted the Hungarians, but seeing their lord was dead (as Anonymus, the notary of King Bela, relates in his "Historia Ducum Hungariae", ch. VI), of their own free will elected as prince Tuhutum, the duke of the Hungarians, and took an oath of allegiance.

Gyula the Elder, grandson of Tuhutum and duke of Transylvania too, went to Constantinople and there converted to the Christian religion of the rite of the Eastern Church and taking along with him the monk Hyeroteus who later became bishop, brought several people of his own over to the same faith, as proved by Samuel Timon in "Imago antiquae Hungariae", book 3, ch. 5.

In the 11th century, after Saint Stephen the King defeated Duke Gyula the Younger, he united Transylvania to the kingdom of Hungary; and all the other Hungarians in Transylvania passed over to the Christian faith; however, through the endeavours of the priests of the Western Church who were working to convert them, they passed over to the Western Church; and as the other Hungarians converted to the religion of Christ already in the time of duke Gyula the Elder had, one after the other, passed over to the Western Church, the Romanian inhabitants of the Province were the only ones to remain in the bosom of the Eastern Church.

Civic rights were common

As this difference in matters of religious rites was in no way a hindrance, civic rights were common to the two peoples, the Hungarians and the Romanians, since the time of Duke Tuhutum when they united into a single society. To make a long story short, we bring as proof of this truth the authentic letter of the Convent of the Blessed Virgin Mary of Cluj-Mănăştur, written in 1437, which the illustrious Pray brought to light in his 7th *Dissertatio Historico-Critica*, § 8, with the following content: "*Paul the Great of Vajda Háza, standard bearer of the community of the inhabitants*

of the country, Hungarians and Romanians, in Transylvania, etc.".
In the same place the illustrious author notes that in this letter of
the Convent, the community of the inhabitants of the country,
Hungarians and Romanians, refers to a charter of the *Holy King
Stephen*, that seems to mention their immunities, which would
offer a new proof that both nations enjoyed the same immunities
and the same civic rights.

Special union of Magyars, Szeklers, and Saxons

There is no doubt that in the same century and in the year when
the Convent wrote that letter, the Hungarian nation established
an entirely special union of mutual help both with the Szeklers,
their conationals, who about the same time had begun to constitute
a separate nation, and with the Saxons, brought to the Province
in the 12th century. This union was renewed in the following year
1438, as is proved by the letter issued, in this respect, by the vice-
voivode of the time, Lorand Lépes.

Still that ⟨union⟩ was in no way detrimental to the civic rights
of the Romanian nation; on the contrary their fate was most
flourishing in that very century when the above mentioned Union
was set up; indeed it was from the ranks of that nation that Joanes
Corvinus of Hunedoara rose to the highest public dignities, at first
in Transylvania and then he rose to be supreme commander and
governor of the kingdom of Hungary, while his immortal son Matthias
occupied even the throne of the country; not to mention other princes
of Transylvania, men risen from this nation, Ioanes Geczy, some
of whose descendants are said to be still living in the county of
Dobîca, held, very honourably, the high dignity of governor of
Transylvania, while Stephen Josika, a Romanian by birth (according
to the historian Wolfgang of Bethlen) held the high office of chancellor
under prince Sigismundus Báthory by the end of the 16th century.
About the middle of the same century Nicolaus Olahus, of Romanian
parents too, born in Sibiu, won fame in Hungary through the dignity
of archbishop of Strigonium and chancellor of the country, while
his brother Matthew held the hereditary post of king's judge of
the *sedes* of Orăștie, in Transylvania. And these illustrious men,

both in Hungary and Transylvania, did not conceal their Romanian origin; on the contrary, according to the evidence of Bonfinius and of Lucius the Dalmatian, the two Corvinus were proud that together with their nation they descended from the Roman colonies. Besides, Emperor Ferdinand testified to the origin of the Corvinus and of the above mentioned archbishop of Strigonium, his chancellor, ⟨sons⟩ of the Romanian nation, the offspring of Romans, when in the diploma granted to the family of the above-mentioned archbishop, on November 23, 1541, he expresses himself as follows: "Thus are almost all the origins of the most illustrious peoples, among which the Romanians, your conationals have not the least illustrious ⟨origin⟩, as it is well known they descend from the masters of the world, from the city of Rome; that is why even now in their language they are named Romans; this people of yours was almighty due to its valour, and gave birth to very many great captains among whom Ioanes Hunyadi, father to the famous King Matthias, and your forefathers, living almost in his time, are said to have distinguished themselves".

After the union established in 1437 and 1438 between the Hungarians, the Szeklers and the Saxons, the Romanian people gave birth to all those great men and to many others who deserved well of the Homeland ⟨and⟩ who were elected and raised by the free suffrage of the Estates and Orders to hold the highest offices and dignities, so that during all the centuries having elapsed until the 17th, the Romanian nation permanently enjoyed all the civic rights, the same as the other citizens, and the above mentioned union never prejudiced in the least the civic rights of the Romanian people, or did not even try to prejudice them.

At that time the Reformation of the Western Church started in the 16th century by Calvin, Luther and Socinus, was making great progress in Transylvania too, and as there occurred a division between Hungarians, Szeklers and Saxons in matters of religion, into four parts, namely, Roman Catholic, Calvinist-Protestant, Evangelic-Lutheran and Socinian-Unitarian, in the public assemblies of the Estates and Orders several articles were drawn up through which measures were taken to ensure the security of these parties disunited as regards religion; and all the four religions of the Hungarians, the Szeklers and the Saxons, divided into four parts in accordance with their diversity, were proclaimed to be accepted by the laws, as is clearly laid down in the *Approbatae Constitutiones* of the country, collected by order and with the authorisation of prince George Rákóczy from the Diet articles, from 1540 to 1633, part I, tit. I, art. 2.

But these articles made no mention of the Greek Eastern Church which the Romanian people had embraced at the moment when they became Christians. But it could not be mentioned since these articles concerned only the religions sprung up in Transylvania after the Reformation of the Latin Church. Consequently the Greek Eastern Church in Transylvania, or better said the Romanian people belonging to it, was left in the same condition as it was before these articles were drawn up, that is to say, with the freest exercise of its religion and the enjoyment of all the rights connected with it, which is fully proved by the charter given by Queen Isabella as concerns the bishopric of Greek rite of Geoagiul de Sus, presented to the worthy Christophorus together with all the appurtenances, incomes and benefits of that bishopric in the year 1557 when the Latin bishopric of Alba was dissolved ⟨and⟩ all its property, appurtenances and incomes were allotted to the fisc, as published by the illustrious Benkö in his work on Milkovia, § 145, lit. C.

> 17th c. harmful alterations in the conditions
> of the Romanians: the *Approbatae Constitutiones*

But all the happy fate of the Romanian people changed in the recent 17th century both as regards its civil condition and the religious one, namely in 1613, 1630 and 1649, after the Hungarians, the Szeklers and Saxons had renewed the union concluded in 1437 and 1438 and were careful to introduce in the collection of laws made ⟨beginning with⟩ 1540 or in the *Approbatae Constitutiones* both the terms of this union, and the articles made in the above mentioned manner in favour of the four religions which they belonged to, though divided in them on this point: indeed in this collection of laws or *Approbatae Constitutiones* the following were introduced ⟨things⟩ very detrimental and harmful to the Romanian nation and religion, which, for centuries, had enjoyed a legal existence in the Province:

Part I, tit. VIII, art. 1: *Though the Romanian nation is not considered in the country among the Estates and its religion is not among the "received" ones, still as long as they are tolerated for the benefit of the country, the clergy shall observe the following, etc.*

Part I, tit. IX, art. 1: *Though the Romanian nation is admitted in the country for the sake of the common good, still, overlooking its lowly condition, etc.*

Part III, tit. I, art. 1: *Considering that the country consists of three nations (observing the laws of the country), if any nation were harmed in its liberties, privileges, ⟨and⟩ customs, it is the duty of the others to, etc.*

Part III, tit. LIII, art. 1: *Since the religion of the Romanian nation is not one of the four "received" religions, neither the monastic order, etc.*

How it was possible to introduce these ⟨things⟩ into the collection of laws is incomprehensible. For, as the preface and the *Approbatae* admit, this collection could be composed and formed only of decrees and Diet articles beginning with the year 1540, when Transylvania was separated from Hungary, until the year 1653. But from 1540 to 1653 there was not a single Diet article, so much the less an older one, which could decide so positively that *the Romanian nation must not be considered as belonging to the Estates nor its religion among the "received" ones, but that both the nation and the religion must be admitted in the Province only for the common good.*

As there was no actual law of this kind, the conclusion must be drawn that it was only through a mistake or the carelessness of the compilers that additions were introduced in the collection; if, however, an intention to do harm exerted any influence, it is not the duty of the petitioning nation to discuss it. Now it is clear that in the collection we come across certain expressions which clearly point to the compilers' hatred rather than love for the Romanian nation. For example the expression which appears in part I, tit. IX, art. 1: *still, as it ⟨the Romanian nation⟩, overlooking its low condition...*

Thus, whether the above mentioned additions were introduced into the collection by mistake or due to carelessness, or that they

⟨were introduced⟩ deliberately to do harm, their introduction was effected warily ⟨and namely in such a way⟩ that the additions be introduced into the *Approbatae* not in direct style and in the form of a legal statute (for it would not have been possible to do it without quoting the article on which they were based, an article which never existed) but must have been put in before, as a preamble to the other legal statutes, inserted in the collection with the conjunctives *though* or *because*.

Remarks against groundless premises

Though such premises usually have no force if they are not supported by a legal basis, they were so successful with many citizens of the Homeland that very soon the opinion that the Romanian nation and religion were only tolerated in the Principality became almost generally accepted and even spread to the foreign nations.

In order to oppose this opinion which is based only on the above premises, which are completely groundless, the following remarks must be made concerning those ⟨premises⟩, ⟨remarks⟩ produced by the very laws and history of the Homeland.

These premises state: *Though the Romanian nation is not considered one of the Estates but is admitted in the country for the sake of the common good.* How far the supposition that the Romanian is not considered one of the Estates is from the truth, is proved by the authentic letter mentioned above sent by the Convent of the Blessed Virgin of Cluj-Mănăștur in 1437, which says plainly: *the community of the inhabitants of the country, Hungarians and Romanians, in Translyvania;* nay this genuine document proves plainly that, at the time when it was issued, the Estates of the Province or the civic community consisted only of the Hungarian and Romanian nations and, as no subsequent law can be mentioned which should have decided positively that the Romanian nation should not be considered one of the Estates and as the inhabitants of the country, it is clear that the assumption of the above mentioned premises is quite groundless.

What the addition that *the Romanian nation is admitted in the country only for the sake of the common good* means, is quite incomprehensible, as the same ought to be said of any other nation; moreover the term *admitted* does not suit the Romanian nation which is much older than all the other nations of the Province; it might be assigned with good reason to the other nations which ⟨as⟩ the history of the Homeland and the diplomas of the princes show, came to the Principality far later than the Romanian nation and were admitted to live and ⟨enjoy⟩ civic rights, either through compacts concluded or through privileges granted by the princes.

What Anonymus says

Both the history of the Homeland and Roman history prove that, for certain, the Romanians had been living in Transylvania many centuries before the Hungarians came to it and when, losing in a battle their Duke Gelu, they no longer opposed the Hungarians, but of their own free will, holding out their right hand, chose the duke of the Hungarians, Tuhutum, to be also their ruler, and by this very fact they admitted the Hungarians to cohabit with them, to enjoy equal citizenship, and common civic rights. The Hungarians were glad of this free and spontaneous action of the Romanians and both nations found in equal citizenship and community of rights their happiness which they were unwilling to risk in a subsequent war whose uncertain end both feared. This sets off clearly the compacts concluded between both nations, mentioned by the words of Anonymus, notary of King Bela: *holding out their right hand, of their own free will, the Romanians elected Tuhutum duke of the Hungarians to be also their ruler.*

Besides history, the privileges and diplomas granted by the princes prove that the Saxons came to Transylvania in the 12th century, ⟨and⟩ the Armenians and Bulgarians in the 17th century and that ⟨they also⟩ obtained admission.

There are also the Germans, citizens of the Homeland, who, as history also proves, came to the Province especially by the end of the 17th century with the army of the blessed Emperor Leopold and obtained admission exactly in the same way as the Hungarians ⟨obtained it⟩ who came by the end of the 9th century. By the end of the 17th century the Estates of Transylvania submitted willingly to the rule of the House of Austria, when there was an imperial army in the Province, as the Romanians, by the end of the 9th century submitted to Tuhutum, duke of the Hungarians who had just come, and elected him, of their own free will, to be also their ruler, holding out their right hand. Thus the same as the Hungarians under Duke Tuhutum were admitted to equal citizenship with the Romanians, the Germans under Emperor Leopold ⟨were admitted⟩ to equal citizenship with the other citizens of Transylvania.

The term "tolerated" is incompatible with both the ancientry and the religion of the Romanians

The above mentioned additions say also: *still as long as they ⟨the nation and religion of the Romanians⟩ are tolerated for the common good of the country, etc.* No doubt the same might be said about the other nations and religions, for they were certainly tolerated for the good of the country; for, if they were not tolerated, there would follow to the great detriment of the Province, either emigration or discord and dangerous disturbances. As the phrase *to be tolerated,* in the present style of the Court means the same as *to be deprived of lawful existence,* it can be applied neither to the nation nor to the religion of the Romanians, since what was said above shows for certain that this nation and this religion were the oldest in the Province and that the former enjoyed civic rights, and the latter free and public exercise and that there is no law which deprived the nation or the religion of the Romanians of its rights, consequently of a lawful existence, and declared them to be tolerated. ⟨This⟩ is not contradicted by the fact that the religion of the Romanians was not mentioned especially among the religions named *received* in art. 2, tit. 1, of part I of the *Approbatae Constitutiones:* indeed, even the old Diet articles, which the above-mentioned law of the *Approbatae* consist of, prove clearly that their object was only the religions which either had not existed in the Province before but penetrated there for the first time in the 16th century, such as the Protestant, the Evangelic Lutheran and the Unitarian religions, or had existed before, but through a public law had seen their free exercise hindered and whose rights possessed before were almost completely wiped out, such as was the case of the Roman Catholic religion. The religion of the Romanians was not one of those, thus it was not the object of the above mentioned articles which the law of the *Approbatae* consisted of and could therefore not be mentioned in that Law: in this real sense there are only four *received religions* and the religion of the Romanians is not among them. But if the word "received" is to be taken in the sense that only the religions named "received" can enjoy a lawful existence and free exercise, then this name can be so much the less denied to the religion of the Romanians, as it is certain it is the oldest in the Province, it has always enjoyed free exercise and was never denied this by any public law.

> On the meaning of:
> *The country consists of three nations*

As for the addition inserted in the conditions of the Union of the three nations in the *Approbatae Constitutiones*, part. III, tit. 1, *as the country consists of three nations*, anyone examining the content of the whole paragraph in which it is inserted will clearly see it has no other meaning except that only three of the nations in the country formed a union. This addition however, can be so much the less taken to mean that only the three united nations constitute the community of the citizens, or the Estates and Orders of Transylvania, as the above-mentioned shows clearly that the Romanian nation enjoyed civic rights long before the union between the Hungarians, the Szeklers and the Saxons, and that it constituted, together with the Hungarian nation, the community of the inhabitants of the country without ever having been despoiled of its rights by any precise law.

Consequently, the present sad fate of the Romanians in Transylvania is not due to the laws but to the iniquity of the times: it will never be possible to prove that the legislative power in Transylvania committed the injustice of depriving and despoiling the oldest nation in the Province of its civic rights sanctioned as shown above by the compacts concluded. Nay, even if an unjust fate had caused things to go so far to deny the clergy of this nation the rights the clergy of the other nations enjoyed and the nobility from posts and dignities, especially the higher ones, this nation was never denied free exercise of its religion, and nobiliary exemptions and privileges: but the Romanian nobles, who having abandoned their Greek religion embraced the Roman Catholic or the Protestant one, or those whose forefathers did so, have been able and are still able to succeed in obtaining the highest dignities. We shall take the liberty of quoting as example the illustrious families of the counts Kendeffi and of barons Josika, Huszar, Nalaczy and a great many noble families native of the Hunedoara county and the Făgăraș district who long ago or more recently passed over to the Roman Catholic or Protestant faiths. These families held and some still hold dignities, even high dignities in the Principality, though most of the citizens of the country know quite well that these families and some others too, among the outstanding ones, are the offsprings of the descendants of the Romans, that is to say of the present people of the Romanians, and their written documents, especially the oldest ones, if examined with a critical eye testify ⟨to this⟩; and seeing that even under the princes who ruled the Principality after the compiling of the *Approbatae Constitutiones*, in which the above mentioned additions were inserted encouraging the opinion that the Romanian nation was only tolerated, many members of the Romanian people were raised to the ranks of real noblemen for merits gained in battles and many were given rights of possession and they all benefit to this day of all the rights implied by the Estate of noblemen and of donees in exactly the same way as the noblemen and donees of the other nations of the Province; this finally proves that the Romanian nation was never despoiled of its civic rights by the legislative power and proclaimed *tolerated;* otherwise the nobles created from among its ranks could have enjoyed only the honour of being noble, as to this day is the case of the Armenians, but not the rights and exemptions the law grants the nobles.

Moreover, such a statute through which a nation, older and more numerous than the others, could be deprived of civic rights would never have been sanctioned by the prince for several reasons, and if it had been made without his authorization it would never have been sanctioned, but would sooner have ruined the civil society founded by the Hungarians and Romanians by the end of the 9th century when they held out their right hand, under Duke Tuhuțum, and would have brought the two nations back to the condition they were in, before the choice made by the Romanians, of their own free will, of the duke of the Hungarians Tuhutum, as ruler of their country, that is to say in a state of war; therefore they would have had to fight one another again until one of the nations subjugated the other, or would have united again through new compacts. But as it cannot be proved that they did either of these things, the compacts concluded before remained in force.

Such being the state of things it is clearly seen that the nation and religion of the Romanians suffered the loss of a part of their civic rights not through a public law, but through the iniquity of the times; the opinion that this nation and its religion are only tolerated in Transylvania is based on the above mentioned additions inserted in the *Approbatae* collection of laws, ⟨but⟩ which are utterly groundless and that it rests only with the spirit of justice and mercy of the Prince to free this nation from so cruel a fate and to reinstate it in all the rights it enjoyed before.

The rights of the Romanians were acknowledged by Leopold I, Maria Theresa, and Joseph II

The late august Prince and Emperor Leopold I acknowledged this; not only did he settle that the clergy of this nation, which by the end of the last century passed over to the Union with the Roman Catholic Church should be admitted to all the rights, privileges and benefits enjoyed by the clergy of the other religions, but also on the strength of the diploma issued on March 19, 1701 he decided that all the Romanians, the laymen and the plebs alike, who would unite with the Roman Catholic Church shall be immediately considered as belonging to the Catholic Estate and, thus, be counted among the Estates and allowed to enjoy, like the other sons of the Homeland, the laws of the country. The august and of blessed memory Princess and Empress Maria Theresa also acknowledged this when, on the occasion of the beginning of her reign she sanctioned the rights of all the inhabitants of the Province and especially the privileges and diplomas bestowed upon the Romanians by her august grandfather.

It was acknowledged finally by that blessed emperor of eternal memory, the all-just and almighty Emperor Joseph the Second who understood the pure and simple rights both of man and of the citizen, who realised the injustice and oppression, who saw with his own eyes and was fully persuaded that the Romanian nation was far more numerous than the others in the Province and of great use, both in times of peace and of war; that is why, in his desire to carry out his task as an all-just prince, and restore to the citizens their rights in order to prevent discord and disagreement between the nations, overlooking all the prejudices of those who were opposed, often decided, with great mercy, that in the future, wiping out completely any unjust discrimination of inequality, the Romanians, even irrespective of their nation and religion, should fully enjoy the same rights and benefits as the other peoples in this Principality, and since they are charged with equal tasks, they should be granted equal rights and benefits.

But these diplomas, decrees and decisions of the August Princes have had so far little effect; indeed the clergy united to the Roman Catholic Church has been, it is true, admitted to enjoy certain exemptions, but not all the rights and benefits the clergy of the other nations enjoy; but the clerics of Greek rite, non-Uniate, though the same duties are incumbent on them as on the clerics of the other religions, are not even exempt of contribution (except for the capitation). Until the present the Romanian nobles, especially those in the counties, obtain with great difficulty certain posts only, and especially unimportant ones. As for the townsfolk or the plebs, they are not given access to the arts or to the handicrafts.

The restrictions imposed by the Diet in 1744

Besides, the Diet held in 1744 by the Estates and Orders of the other nations, stated in art. VI. that the above mentioned concessions made by the blessed Emperor Leopold I and granted to the Romanians, laymen and plebs, and confirmed by the blessed Empress Maria Theresa *referred only to the ecclesiastics and to those enjoying nobiliary prerogatives, whose condition is the same as that of the other citizens of the Homeland, and who are considered as belong-*

ing to that nation of the three constituting the system of the Principality where they have settled through the acquiring of property without by this fact, forming a fourth nationality, but that they cannot and must not be extended to the plebs, for fear the system of the Principality be disturbed and for fear the Romanian plebs and that of other aliens be considered as a nation and be even detrimental to one of the three nations and to their rights, privileges, exemptions and prerogatives.

The supplicating nation recognizes, in fact, that the condition of the Romanian nobles, as regards the use of nobiliary privileges, is the same as that of the other citizens of the Homeland, but, it is sorry to have to admit that not a single person who keeps his Greek rite and religion ⟨is admitted⟩ to higher dignities, and very few are admitted even to smaller posts. And even so with great difficulty. It must further admit it knows nothing of the law on the basis of which the Romanian nobles were considered as belonging to the one of the three "received" nations where they settled through acquisition of property, — all those who have a thorough knowledge of the history of the Homeland know perfectly well that the Romanians settled and fixed their abode in Transylvania a few centuries before the three "received" nations, and by their own free will, holding out their right hand elected as their ruler, by the end of the 9th century, Tuhutum, duke of the Hungarians, who had entered that Province then and that ever since that time the community of the inhabitants of Transylvania was constituted of Hungarians and Romanians, but not that the latter were considered among the former or among other nations. The supplicating nation ⟨must admit⟩ further that it knows nothing of the law on the strength of which one nation or another of Transylvania was declared *received;* it is common knowledge that the term "received" began to be used first in the 16th century with regard to the new religions sprung up after the Reformation and that according to the laws enacted then, a "received" religion meant nothing but a religion allowed free exercise, but as for one nation or another, this term, according to the laws preceding the statement made by the Estates in 1744, was not used anywhere, and the nations enjoying lawful existence are called *civic* and the above proved clearly that the Romanian ⟨nation⟩ was counted among them. It is also certain that in Transylvania there are *united nations* and that there are three that constitute this union, and that the Romanian ⟨nation⟩ is not counted among them. That is why, since that union could not be at all effected in a way that could be detrimental to the civic rights of the Romanian nation, in fact the oldest of all the nations in Transylvania, or abolish them, it was a heavy and terrible blow to this nation when the above-mentioned statement of the Estates in 1744 put it on the same footing as the aliens. Not alien, but old and even older by far than all the other nations is the nation of the Romanians in Transylvania; it never demanded and does not demand the overthrow of the system of the Principality, but rather its restoration, when it demands to be reinstated in the civil and civic rights which no law dispossessed it of, but the iniquity of the times. This reinstatement shall never harm any of the three nations or their rights, privileges, exemptions and prerogatives, since through it the nation will obtain nothing but what it possessed before and lost due to the iniquity of the times; it shall never be necessary to set up a fourth nationality for the supplicating nation, as for many centuries it has constituted, immediately after the Hungarians, the second civic nation.

The evil effects of the statement made in 1744

And the fact that in the above mentioned statement made by the Estates in 1744 *only* the Romanian nobles are granted the same condition as the other *citizens* of the country, while the Romanian plebs is excluded from it, was indeed a heavy blow to the supplicating nation and to its ⟨deep⟩ regret and to that of the Province it bears the evil effects of that statement. Indeed, to pass certain things over in silence, since then certain circles on the Königsboden have tried twice to drive away that unfortunate plebs, though, besides what was said above, even the basic privilege which mentions the Königsboden granted to the Saxon nation clearly proves that the Romanians there must enjoy the same rights and liberties as the other nations.

Appendices 133

The supplicant nation desires that its plebs
be treated in the same way as the other plebes

It is certainly not the intention of the supplicating nation to obtain for its plebs more than it is legally entitled to; but as the above reveal plainly that it is entitled to all that the plebs of the other nations living in the Principality is entitled to, and ⟨as⟩ this ⟨latter⟩ is not excluded from the number of the Homeland's citizens, this is the one thing the supplicating nation desires: that its plebs be treated in the same way as the plebs of the other nations and, moreover, since it bears the same civil charges as them, it should enjoy the same benefits, in fact, as required by several all-high orders.

Seeing that these orders and the above-mentioned diplomas of the august predecessors of Your Holy Majesty, granted in favour of and for the improvement of the condition of the Romanian people here had no effect or but a slight effect, or a short-lasting effect, but invariably an uncertain one, and in consequence of which the supplicating nation has not been reinstated in the common enjoyment of the rights of the civil society, which a cruel fate deprived it of, and seeing that until this day it has been compelled to bear only obligations and be deprived of civic benefits (and this is not only contrary to the laws of justice and fairness, but causes great prejudice to the public state), considering that, as long as its clergy and nobility are maintained in a lowly state, this nation cannot hope to own a culture of its own and therefore an increase of ⟨its⟩ diligence, on the contrary we fear that ignorance, laziness and sloth, accompanied by all the vices that usually spring from it then will develop amid it, causing disorder in the Province, and that, besides, mutual mistrust between this nation and the others, internal hatred bred here, as well as the unrest and sufferings of these unhappy souls will increase more and more; they at the same time shall endanger the security and public and private peace (which all these things, many citizens of the Homeland wish to prevent, thinking of justice and equity, for a long time have demanded that the supplicating nation be reinstated in all ⟨its⟩ civic rights):

The Romanian nation's requests

Consequently,

The Romanian nation coming most humbly before the throne of Your Majesty, with all due respect and submission prays beseechingly the following:

1. That hateful and insulting terms such as *tolerated, admitted, not counted among the Estates* and others of the same kind which like external stains have been unjustly and unlawfully stamped ⟨on the forehead⟩ of the Romanian nation, be completely removed, revoked and abolished publicly, as unjust and shameful; thus due to the mercy of your holy Majesty the Romanian nation, reborn, shall be reinstated in the enjoyment of all the civil and civic rights.

Consequently,

2. The supplicating nation should be replaced among the civic nations on the same place as it held, according to the above mentioned evidence provided by the Convent of the Blessed Virgin Mary of Cluj-Mănăştur in 1437.

3. The clergy of this nation, faithful to the Eastern Church, no matter whether it holds the same opinion as the Western Church, the nobility and the plebs, both in towns and in the countryside should be considered and treated fairly, like the clergy, the nobility and the plebs of the nations forming the system of the union, and be made to share the same benefits.

4. On the occasion of the election in counties, *sedes*, districts and urban communities of clerks and deputies to the Diet, and when new appointments happen to be made or promotion given to posts in the aulic and provincial dicasteries, a proportional number of persons of this nation should be appointed to these posts.

5. The counties, *sedes*, districts, and urban communities in which the Romanians exceed in number the other nations, should bear also Romanian names while those in which the other nations form the majority should be named after them or bear a mixed name, *a Hungarian-Romanian, Saxon-Romanian name*, or, by completely removing the name taken after one nation or another, the counties, *sedes* and districts should preserve the name they have had down to the present after rivers or fortresses and it should be stated that all the inhabi-

tants of the Principality, irrespective of nation or religion should use and enjoy, depending on the Estate or condition ⟨of every one⟩ the same liberties and benefits and bear the same obligations, to the best of ⟨their⟩ ability.

Recapitulation of the grounds of those requests

All that has been said above fully proves that these demands are based on natural justice and the principles of the civil society as well as on the compacts concluded. And considering that, as early as 1761, according to the census taken in all circles and places (except for the district of Brașov) the supplicating nation amounted to 547,000 people, and (if the Romanians in this district at the time were only 13,000) this nation numbered 560,000 persons, while all the other nations all together (including the Romanians who embraced another religion) according to the census taken in 1766 numbered only 392,000 and a few hundred persons, and ⟨since⟩ this leads to the conclusion that the present population of Transylvania, according to the 1787 census, consists of one million and about seven hundred thousand people, the most considerable part of which, maybe even a whole million, consists of the members of the supplicating nation while, besides, two whole frontier regiments in the Principality and more than two thirds of the three field regiments and more than one third of the Szekler cavalry regiment consist ⟨also⟩ of men from the supplicating nation, and, ⟨as⟩ in general the public obligations of the Province are borne by the supplicating nation, in greater amount than by all the other nations taken together, proportionately to ⟨its⟩ larger number, it places its trust in the paternal love of Your Majesty to see that its demands, being perfectly just, will have the expected effect, especially as it depends only on the supreme authority of Your Majesty to reinstate it in the enjoyment of its civic rights which it has been deprived of not by law, but by an unjust fate.

If requests are not satisfied, the supplicant nation
asks for a national assembly of her own

Still, if it is the intention of Your Holy Majesty that these requests, made by a nation always loyal to the August House, be communicated first to the Estates and Orders now assembled in the Diet here, in the Province, through the royal plenipotentiary commissioner, in this case it might easily happen that, despite the justness of the request, fully proved by the above, and ⟨despite⟩ the wish of a great many inhabitants of the country assembled in the Diet who have no other desire but the happiness of ⟨our⟩ beloved Homeland, and the promotion of the public good, ⟨wishing⟩ to satisfy the request of the supplicating nation, still some of the citizens of the Homeland now present at the Diet, either referring to some usage opposed to the rights of the supplicating nation, ⟨a usage⟩ confirmed maybe by some prescription (though there can be no prescription regarding the civil society) or not knowing well enough the rights of the civil society, or, finally, not having studied enough the history of the Homeland and the sense of the laws of the Homeland and not realising clearly enough the justice of the rights of the undersigned nation, and maybe partly prompted by a secret national and religious hatred, might attempt to prevent and check, at a certain moment, the happy result of this thing, the undersigned nation most humbly begs Your Holy Majesty to kindly and mercifully permit, in this case, that, in a national assembly, whose modality and place will be settled by the two bishops of the nation in the Province, taking counsel with a few members of the clergy, the nobility and the military Estate, in order that as soon as possible proposals might be set before Your Holy Majesty, a number of deputies be chosen and empowered to plead and defend the cause of the Romanian nation wherever necessary; should obstacles be put in the nation's way, despite its rights, to draw up orderly all the grievances of this tortured nation and set them before Your Holy Majesty to be mercifully remedied.

The supplicant nation — almost a whole million people

Your Holy Majesty kindly imparted, with boundless mercy, such a grace and comfort not only to that part of the supplicating nation living in the Banat and in the counties of Hungary in the vicinity of the Banat and Transylvania, but to the whole Illyrian nation and to all the nations in the vast Monarchy; they all presented in public assemblies, to Your Holy Majesty, their grievances and requests and none were left uncomforted from the august throne of mercy; that is why the petitioning nation, comprising almost a whole million people, and who, it is true, lives at the furthest away border of the Monarchy, but with hearts and souls always loyal to the august House of Your Majesty, hopes that it will obtain from the fount of justice and mercy, the comfort it is beseeching.

Your Holy Majesty's
Most humble for ever loyal subjects,
the Clergy, the Nobility, the Military, and the Urban Estates of the Whole Romanian Nation in Transylvania.

D. Prodan, *Supplex Libellus Valachorum or The Political Struggle of the Romanians in Transylvania during the 18th Century*, Bucharest, 1971, Annexes, pp. 455—466 (Translated by Mary Lăzărescu).

APPENDIX C

1842

Stephan Ludwig Roth
Der Sprachkampf in Siebenbürgen. Eine Beleuchtung des Woher und Wohin?

> "Denn sie säen Wind und werden Ungewitter einernten" (Hosea, VIII, 7).

Die Herrn auf dem Landtage in Clausenburg mögen eine C a n z l e i s p r a c h e gebäret haben, und sich nun freuen, daß das Kind zur Welt gebracht ist — eine Sprache zur Landessprache zu erklären, hat nicht Noth. Denn eine L a n d e s s p r a c h e haben wir schon. Es ist nicht die Deutsche, aber auch nicht die Madjarische, sondern die W a l a c h i s c h e ! Mögen wir ständische Nationen uns stellen und gebehrden, wie wir wollen, es ist nun einmal so, und nicht anders. Pst, Pst! sagt man, und zupft mich am Aermel:"Einfältiger Kerl, so etwas sagt man ja nicht"! — Diesen Ehrentitel mag ich vielleicht verdienen, auch um meiner andern Streiche willen — aber hier grade, scheint mir, belohnte man mich über Verdienst. Denn ich und du und er, wir, ihr, sie alle haben diese Ueberzeugung. Wenn man von einer allgemeinen Sprache des Landes redet, glauben wir, daß damit keine andere gemeint sein könne, als die Walachische. Umsonst stecket der gejagte Strauß seinen Kopf in den Strauch, der Meinung, weil er nicht sehe, würde auch er nicht gesehen. Umsonst, meine ich, sagt man so etwas nicht; wenn man's auch nicht sagt, ist es deßwegen doch. Lieber gesagt und darüber gedacht als nicht gesagt und nicht gedacht. Es ist diese Thatsache nicht zu leugnen. Sobald zwei verschiedene Nationsgenossen zusammenkommen, die ihre Sprache nicht können, ist gleich das Walachische, als dritter Mann, zum Dolmetschen da. Man mache eine Reise, man begebe sich auf einen Jahrmarkt, Walachisch kann Jedermann. Ehe man den Versuch macht, ob dieser deutsch, oder jener madjarisch kann, beginnt die Unterredung in walachischer Sprache. Mit dem Walachen kann man ohnedem nicht anders reden, denn gewöhnlich redet er einzig die seinige. Das kommt daher: Um madjarisch oder deutsch zu lernen, bedarf man des Unterrichts und der Schule; walachisch lernt man auf der Gasse — im täglichen Verkehre — von selbst. Die Leichtigkeit ihrer Erlernung beruht nicht nur in der großen Menge lateinischer Wörter, welche dieses Mischlingsvolk, durch die Verschmelzung mit römischen Colonisten, in sich aufnahm, und welche uns Siebenbürgern, bei unserer bisherigen lateinischen Erziehung, von selbst verständlich sind — sondern das Leben selbst bringt uns alle Tage in Verkehr mit diesem zahlreichen Volke, welches beinahe die Hälfte der gesamten Bevölkerung bildet. Heute bleibt ein Wort hängen, morgen das andere, und nach einiger Zeit bemerkt man, daß man walachisch kann, ohne es eigentlich gelernt zu haben. Würde es einem aber auch nicht so leicht, so empfiehlt deren Erlernung ein tausendfältiges Bedürfniß. Will man mit einem Walachen reden, so muß man sich zu seiner Sprache bequemen, oder man halte sich gefaßt auf sein achselzuckendes: *Nu știu!*

St. Ludwig Roth, *Der Sprachkampf in Siebenbürgen. Eine Beleuchtung des Woher und Wohin?*, Kronstadt, Druck u. Verlag von Johann Gött, 1842, p. 47—48 (Ch. VI: *Panslavismus, oder Walachen und Adel*); see also Stephan Ludwig Roth, *Viața și opera*, Ed. Carol Göllner, București, 1966, p. 270—271.

APPENDIX D

1891

A. Nicolson, Consul-General of Great Britain at Budapest
Report on the Political Situation in Transylvania

(Confidential)

The land and her population

AMONG the several States or provinces of which the Austrian Empire was formerly composed, Transylvania ranked third in extent, surpassed by Hungary and Galicia, but of greater size than Bohemia, Styria, or Dalmatia. Her population numbers at the present day about 2,100,000 — of whom 1,200,000 are Wallachs or Roumans; 700,000 Magyars and Szeklers; and 200,000 Germans, or, as they are termed, Saxons. The Szeklers are supposed to be the descendants of the earliest Hunnish invaders; but they are now practically for all purposes identical with the Magyars, and are chiefly settled in the extreme east of Transylvania, on the frontiers of Roumania. Beyond, and bordering on the western confines of Transylvania, in Hungary proper, are districts containing a large Rouman population to the number of about 1,500,000 who had migrated over the frontier from their native homes, so that the total Rouman population in Hungary at the present day is, at the lowest estimate about 2,600,000; all living in fairly compact masses, knit together by one language and common political aims, and forming, owing to their geographical position and racial affinities across the frontiers, an important and solid factor in the future political development of the Hungarian Monarchy. As to whether the Roumans or the Magyars were the prior occupants of the country is still a point in great dispute; but as this is a matter which has little bearing on the questions with which I propose to deal in this Report, it may be left as an undecided, and probably insoluble, problem.

An historical sketch of the early relations between Transylvania and Hungary

Before entering upon the subject of the present situation in Transylvania, it is necessary to give a very brief historical sketch of the former relations between that province and Hungary. Until the year 1526, Transylvania was governed by her own Dukes or Voïvodes, who were vassals to the Hungarian King, but who otherwise enjoyed considerable independent authority. In 1526, the Hungarians were completely overwhelmed and defeated by the Turks at the battle of Mohacs; and a few years afterwards, in 1538, a Treaty was made between Ferdinand I of Hungary and John Zapolya, Prince of Transylvania, by the terms of which the latter country was completely separated from Hungary, on the condition that, on the extinction of the male line of Zapolya, the Princely Crown should revert to the Hungarian dynasty. In 1571* this contingency occurred, as the male line of Zapolya expired, but the condition above mentioned was not fulfilled. Transylvania preferred te become tributary to the Ottoman Sultans while guarding her autonomy, and being governed by her own Princes duly elected by the three "States" or "nations" of Transylvania. These three "nations" were the Hungarians, Szeklers, and Saxons. The Wallachs, or Roumans as they should now be called, who even then formed the large majority of the population, were the serfs, the hewers of wood and drawers of water, without civil or political rights, oppressed by the Hungarian landlord, and despised by the Saxon burgher. They took no part in the political history of the next three centuries, though at times, when driven to despair by exactions and ill-treatment, they rose in sanguinary revolts. In 1691, after the withdrawal of the Turks, the "nations" of Transylvania transferred the sovereignty over their land to the House of Hapsburg, whose chief still bears the title of Grand Prince of Transylvania. The "nations" secured, under an Imperial diploma of the 4th October, 1691, the right of electing all their own dignitaries and officials, and of being governed by their own laws. In this manner the personal union of Transylvania with Hungary was renewed in the person of the Emperor Leopold I,

---

* Clerical error instead of 1541.

who was also King of Hungary. The autonomy conferred by the above-mentioned diploma was confirmed by Charles VI in the Pragmatic Sanction, and by Maria Theresa in 1744. Although throughout many years negotiations for a union between Hungary and Transylvania were continually passing, the Court at Vienna seemed little disposed to encourage these advances; and the autonomy of the province was maintained till 1848. In that year the Hungarian majority of a Diet, assembled at Klausenburg in Transylvania, succeeded in passing a formal declaration in favour of an union with Hungary, the Saxon members protesting; but no effect was given to the vote, as the events of the Hungarian Revolution entirely swept all other considerations from the public mind.

A bright interlude prior to 1868

After the suppression of the Hungarian Revolution, Transylvania, in common with other portions of the Monarchy, was governed under the absolutist system, until in 1861 an era of modified Constitutional Government was entered upon. Diets were called together, in which for the first time the Roumans were recognized as having a political existence, and were allowed to send Deputies; and though the Hungarian inhabitants of Transylvania disapproved of this new order of things, and held aloof from many of the sittings, the Diets enacted some legislation of a useful and liberal character. Moreover, both Roumans and Saxons sent Deputies to the Reichsrath at Vienna; and the few years prior to 1868 are still regarded as a bright interlude in the political history of the former. During that period many Rouman schools were established, the Churches, both Greek Oriental and Greek Roman, were recognized as official creeds, and a great advance was made, which has been steadily maintained, in general enlightenment and education.

In 1868: "the fusion of Transylvania into the newly-established Hungarian Monarchy..."

In 1868, ceding to the insistance of the Hungarian Government, the Court of Vienna sanctioned the fusion of Transylvania into the newly-established Hungarian Monarchy. The Hungarian inhabitants of Transylvania were well pleased with this measure, as they had watched with jealous and mistrusting eyes the emancipation and progress of their former serfs, and could hardly have brought themselves to co-operate, in an autonomous Government, with those whom they regarded as little more than mere uncouth peasants. Both Roumans and Saxons were grievously afflicted by the decision of the Court of Vienna, and expressed in remonstrances and petition to the Imperial Throne their desire not to be handed over to Magyar rule, recalling the many promises, which had been made to them especially since 1848, that their autonomy would be respected, and that no union with Hungary would be sanctioned by their Imperial Master. All was in vain; the Hungarian Government were imperative; and, by a Law of 1868, Transylvania was completely merged in Hungary, was divided into several counties, and was, as far as possible, blotted out of the political Map as a distinctive national body. Roumans and Saxons had perforce to recognize the new situation, and after the first disappointment had passed away, they began to cherish hopes that their lot under their new masters would not be so insupportable. They trusted that the Magyars who had known, by bitter experience, oppression and absolutism, who had fought for the cause of liberty and freedom, would deal liberally and generously with the nationalities who now stood to them in much the same relations they had formerly held towards Austria. Moreover, a law regarding the nationalities had just been passed by the Hungarian Parliament, which seemed to afford every needful guarantee to the nationalities in respect to religion, education, language, local administration, and other matters closely affecting the life of the individual and of the community. Both Roumans and Saxons, therefore, sent Deputies to the Hungarian Parliament, and were ready to act as loyal supporters to the Government at Pesth. But as time passed on, and the Hungarian Government felt that they had freer hands, the present policy of what is popularly termed "Magyarizing" the nationalities was gradually adopted, and was pursued with no greater energy and vigour than in Transylvania. The Roumans, finding their hopes disappointed, altered their attitude also; and a situation has now been evoked, which is not without its elements of future danger.

The serious and specific grievances of the Romanians

During the tour which I have recently made in Transylvania, I found among all intelligent classes of Roumans a feeling of widespread discontent, grounded on more serious and specific grievances than those which exist among the Serb and Croat nationalities, and to which expression and form has been given by a well-organized system of opposition. Furthermore, and this is peculiar to the situation in Transylvania, there is not only openly expressed sympathy in Roumania with the grievances of their brethren, but means have of late been adopted in that country also towards assisting in their alleviation. The students of the University of Bucharest, in the early part of this year, issued an appeal, in various languages, to Europe on behalf of the Roumans resident in Hungary; and a league for the "Intellectual Unity of the Roumans" has been formed in Roumania, under the presidency of the Rector of the Bucharest University, and with branches in various towns in that country.

The Memorandum of the students was intended to enlighten Europe on the condition of affairs in Transylvania; and was, in fact, a long and somewhat bitter indictment against the Hungarian Government; while the Roumanian league, in its published programme, does not conceal the objects of its creation, and specifies, in forcible terms, the injustice and oppressions which it will be its aim to assist in removing. When, therefore, the geographical position of the Roumans in Hungary and their numbers are taken into consideration, and regard is had to the moral support they receive from considerable classes of their brethren over the borders, it is, I venture to think, of some importance to inquire into the actual political situation, and into the causes which have led to its present unsatisfactory condition.

The religious questions

I should like, in the first place, to make a few remarks on the religious questions which, in Transylvania, play a great part, and are closely identified with the national life.

There are two confessions to which the Roumans in Hungary belong; the Greek Oriental Church and the Greek Roman Church, or as they are usually called, the non-Uniates and the Uniates. About 1,600,000 of the Rouman population belong to the Greek Oriental Church, and about 1,100,000 to the Greek Roman Church. The Greek Oriental Church has a Metropolitan, with his seat at Hermannstadt, and two Bishoprics at Arad and Karansebes; while the Greek Roman Church has an Archbishop at Blasendorf, with three Bishoprics at Szamos-Ujvar, Grosswardein, and Lagos [Lugos]. This latter Church is by far the richer and better endowed of the two, as it was much protected and favoured by the Austrian Government in past times, and received considerable endowments and estates, especially under Maria Theresa. On the other hand, the Greek Oriental Church was neglected, and indeed ignored; and stood, until within recent times, under the supremacy of the Servian Metropolitan in South Hungary.

In the year 1846, Shaguna was appointed Vicar-General of the Greek Oriental Church in Transylvania, and from that date a new era was opened. This prelate, who is considered by the Roumans to be one of their greatest men, immediately set to work to revive the Church, which was in a deplorable condition; and the results of his efforts were not only that many converts joined from the ranks of the Uniates, but that he eventually succeeded, in 1868, in procuring the recognition of the Greek Oriental Church in Transylvania as an independent and separate Church, with its own Metropolitan and Constitution. The dogmas of the Greek Orthodox Church in general are accepted; but the liturgy is in the Roumanian tongue; the Bishops are elected by the lay and clerical members of the Diocesan Synods, and the Metropolitan is chosen by the General Assembly. These elections have subsequently to be ratified by the Crown, whose sanction has not invariably been given.

While I was at Hermannstadt, the General Synod was holding its meetings; and I attended one of the sittings, which took place in the small Greek Church which has to serve as a Cathedral. There were forty lay and twenty clerical members present, elected from the various districts in Transylvania and Hungary, and the proceedings were conducted with great decorum and solemnity. The business to be transacted chiefly related to educational establishments, religious funds, pensions, &c. I called on the Metropolitan,

Mgr. Miron, a man of over 60, and apparently rather broken in health. He is a Privy Councillor, and an *ex-officio* member of the House of Magnates, and is generally considered to discharge his duties with credit and success, though several of the more advanced politicians complain that he is too subservient to the Government. This charge, if charge it may be termed, is hardly just, as whenever an occasion has arisen, Mgr. Miron has never hesitated to speak out bravely on behalf of the interests of his flock. His position naturally imposes on him considerable circumspection and tact. The prelates and the priesthood are but poorly endowed; but great attention and care are now being bestowed on the proper education of the village Popes, and no candidate is consecrated who has not passed through a seminary, and received an education suitable to his office. There are still several of the old order still remaining, illiterate and ignorant, and, at some vespers which I attended in a small Roumanian village, I found the officiating priest to be an old man, verging on 80, clad in the peasant's dress, and with merely a stole thrown over his tunic to indicate his ecclesiastical character. The new order of priests are certainly on a much higher level, and probably not much inferior to the average Roman Catholic parish priest.

The Greek Roman Church is more advanced in culture and education than the Greek Oriental, and is, as I have stated, more richly endowed. The Greek Roman Church admits the supremacy of the Pope; the descent of the Holy Ghost from the Father and the Son is accepted, and so is the existence of purgatory; but the Liturgy is in the Roumanian tongue, and marriage of priests is customary. The Rouman is religious; much under the influence of the priests, and strictly observant of his duties towards the Church; but he is, so all authorities admit, exceedingly tolerant towards all other confessions.

Owing to the strife which is now being waged over the confessional schools, both Churches are greatly irritated against the Hungarian Government; and I was told that indirect efforts were also being made by the Government to interfere with Church organization and ecclesiastical funds. The cases cited to me were rather vague; and it would require a very careful inquiry to arrive at the truth in the matter; and for such an inquiry I had neither the materials nor

Appendices

the time. In any case, the priests are at one with the rest of the Roumans in their dissatisfaction, and the members of the two confessions, though differing on religious matters, are united on all political questions. The best example of this is to be found in the fact, that two of the most prominent opponents to the Hungarian policy, the veteran, G. Baritin, [Baritiu] and the editor of the "Roumanian Review", are both loyal and devoted members of the Greek Roman Church.

The school question

The school question is among the chief grievances of the Rouman population; and as it is through the schools that the Hungarian Government are mainly endeavouring to impose their language and national character on the several races, it is round this question perhaps that the fiercest struggle is being waged. At considerable pecuniary sacrifices the Roumans have succeeded in maintaining four gymnasia (combined with a "real" school and two ecclesiastical seminaries), and about 2,500 elementary schools. All of the former institutions and the greater part of the latter establishments were created before 1868, while Transylvania was an autonomous province, under the direct control of Vienna. In 1879, the first step was taken by the Hungarian Government towards what is termed the "Magyarization" of the confessional elementary schools, by decreeing that henceforward the Hungarian language should form an obligatory portion of the studies; and, by a later Law in 1882, it was ordered that no teacher should be considered as duly qualified who had not a competent knowledge of the Hungarian tongue. Out of the 17,000 elementary schools throughout Hungary, under 800 are maintained by the State; and the vast majority of the remainder are supported by the various religious confessions. The Roumans consider that, as the State contributes nothing to the support of these schools, the confessions should be allowed to have a fairly free field as to the instruction they give, provided that the teaching is moral and satisfactory, and not antagonistic to the welfare of the State. Moreover, they point out the extreme difficulty in many districts of finding teachers who have studied Hungarian sufficiently to satisfy the requirements of the Government; and they have had cause to fear that a non-compliance with the regulations of the Government will lead to the closing of the confessional school, and its substitution by a State establishment. The Roumans further complain that the applications which they have made for permission to open one or two more middle schools have been ignored; and that since 1868 no new schools of this character have been established. They also assert that a fifth gymnasium which they possessed has been recently closed by the Government, owing to some non-conformity with the regulations. The Roumans appeal to paragraph 17 of the Law of 1868 regarding the nationalities, in which it is decreed that "the Minister for Public Education is bound to see that, as far as possible, the citizens of evey nationality, if they are living together in considerable numbers, should be able to educate themselves in their mother tongue until they are ready for the higher academical education".

Results of the various attacks on the schools

In a pamphlet which was given to me, it is remarked on this point that "more than two decades have elapsed since the promulgation of this Law, and the Hungarian Government have during that period completely neglected their duty, as they not only have not established a single institution which would enable the citizens to educate themselves in their mother tongue, but they have declined to entertain the proposals of the Roumans, to open some additional middle schools at their own expense".

A case which has excited considerable agitation among the Roumans has been the recent Decree of the Minister for Public Education in regard to the Rouman gymnasium at Belenzes *, an institution which was erected and supported by the Greek Roman Consistory at Grosswardein.

The Minister has ordered that in future Hungarian is to be the language in which instruction is to be given, and that lectures are only to be delivered in the Roumanian language on subjects relating to religion and Roumanian literature.

* Clerical error, instead of Belenyes (Romanian: Beiuș).

It is considered that this Decree will necessitate the closing of the establishment.

I found complete unanimity among both Saxons and Roumans that these various attacks on the schools will have no other result than incense the population; and the endeavour to denationalize either the Saxon or the Rouman by obligatory instruction in Hungarian will only end in failure.

It is, they point out, impossible that an instruction as unwillingly given as it is indifferently received will make any deep impression on the peasant mind, who, on quitting the elementary school, will rapidly forget the smattering he has obtained.

The opinions and characters of the intelligent classes will be even less affected by a compulsory tuition in Hungarian. Living in a polyglot country, they are all good linguists, and would probably in any case have voluntarily learnt the State language. The final touch to the education both of Saxons and Roumans is in nearly every case given either at Vienna, or in Germany, or in Roumania. Many attend the Pesth and Klausenburg Universities as, in order to obtain a lawyer's diploma, a degree at one of those two Universities is necessary; but I was told that the residence of the Rouman in those seats of learning and the treatment they there receive rather increases than otherwise their antipathy to their Magyar Rulers. I visited one or two Rouman gymnasia, and also a superior girls' school, supported by the Rouman community; and all establishments seemed to be maintained in excellent order.

The Transylvanian Literary Association (ASTRA)

There is a Transylvanian Literary Association with its head-quarters at Hermannstadt, which has been established for the purpose of furthering education among the Roumans living in Hungary, and as it is possessed of considerable funds, it has succeeded in obtaining, within the somewhat narrow limits allowed by the State, very creditable results. There are now nearly 3,000 pupils at the middle schools maintained by the Roumans; about 150 students at the Hungarian Universities, and about 100 at the Vienna University, besides several in Germany, Roumania and Belgium.

Growth of an intelligent proletariat

This annual contingent of educated persons will become a good leaven among the population; and it can no longer be said that the Roumans are a race of merely boorish peasants. There, is, however, the danger of the growth of an intelligent proletariat among the Roumans; as, since the service of the State is practically debarred or rendered exceedingly difficult to the educated classes, they either join the already overcrowded ranks of lawyers and doctors, or become journalists. Many now migrate across the borders to Roumania, where they can more easily find a living, and where they are warmly welcomed.

A few are beginning to turn their attention to commercial pursuits; and I was told by Saxons that the Rouman is, as a rule, an exceedingly good man of business, with a great capacity for work. Personally, I was much struck with the intelligence of those with whom I came in contact, and especially with their general knowledge and openness of mind.

Press prosecutions

Another grievance among the Roumans is that in matters of press prosecutions, old laws passed in the time of Austrian absolutism are maintained in vigour for Transylvania alone. Press prosecutions appear to be fairly frequent; and the majority of cases are tried under an Imperial patent of 1852, which is of a far more severe character than any law existing in Hungary. Another grievance is that the cases are universally tried at Klausenburg, the town in Transylvania where the population is chiefly Magyar, and by juries who are of the Magyar race, and, therefore, it is said, prejudiced against the accused. The cases were previously tried at Hermannstadt before Saxon juries, but as the result was generally an acquittal, the venue has now been changed. There were, I was told, four or five journalists now in prison for terms varying from six months to a year. I might, perhaps, be allowed to instance one or two cases. Among the publications is an admirable little monthly periodical, called the "Roumanian Review"*, containing much useful information, and well edited. It is published in the German language, and its object is to bring before the public of Europe the condition of affairs in Transylvania. I have read through many of the numbers. And though the political articles are rather bitter in tone, there appears to be little real dangerous matter in the publication. This Review originally appeared in Pesth; and the editor was eventually prosecuted for inciting the Rouman population to discontent and disloyalty. The editor pointed out that his Review was published in the German language, of which the Rouman peasants were ignorant, and that only a few hundred copies were issued, of which over two-thirds were for foreign circulation. He was condemned to a year's imprisonment and 50 l. fine; and has since removed his editorial office to Vienna, whence the publication is continued undisturbed by any interference on the part of the Austrian authorities.

Another editor, whom I met at Kronstadt, had just completed a term of imprisonment under the patent of 1852; and he informed me that, on his release, he had been met by a large crowd and many mounted men. All of the latter who were recognized were each fined ten florins by the Police authorities for their participation in the welcome.

Another grievance in regard to the press laws is that, contrary to precedent in similar cases in Hungary proper, the defendants, whatever may be the verdict, have to bear the costs of the action. I questioned the "Obergespan" (Prefect) of Kronstadt on the subject, and he assured me that the Hungarian Government were exceedingly patient, but that the limits of decorum were so frequently overstepped by the Rouman journalists, that prosecution were absolutely necessary. He further informed me that nearly all the journalists received pecuniary support from the Roumanians over the frontier; and this intelligence was confirmed to me in a moment of inadvertence by the editor of the "Roumanian Review", who stated that the Roumanian Government subscribed for a considerable number of copies of his publication. If it be true that assistance is obtained from outside, and that this fact is known to the Hungarian

---

* "Romänische Revue", 1885—1894, Editor: Dr. Corneliu Diaconovich.

authorities, it may account, in great measure, for the severity which is exercised against the Transylvanian press. At the same time, it is a question whether it would not be wiser, except in very extreme cases, to diminish the opportunities which are afforded to journalists of acquiring popularity through press prosecutions, and, in view of the small number of the reading public, to allow greater liberty to the press. The paper with the largest circulation issues but little over 2,000 copies daily. In any case, it would be prudent to place the press laws for Transylvania on exactly the same footing as those in other parts of Hungary. It is but fair to state that the tone of the press of Buda-Pesth whenever it has occasion to treat of Transylvanian affairs is not of a character likely to moderate or appease the Rouman journalists; and that most of the incriminating articles in the Transylvanian press have been merely replies to violent attacks on the part of the journals of the capital.

The Law regarding elections, one of the principal grievances

The Law regarding elections is one of the principal grievances of the Rouman population, as owing to it they find themselves unable to secure a fair and adequate representation in Parliament. They have of late years, and in this action they have recently been joined by all Roumans in all parts of Hungary, decided to abstain altogether from participating in the elections as a protest against the exceptional disabilities under which they labour. The franchise is not only, in view of the backward state of the country and the comparative poverty of the people, relatively, but also absolutely, higher than that existing in Hungary. In that country direct taxes to the amount of 5 fl (8s. 4d.) admit to electoral rights, while in Transylvania the suffrage is 8 fl (13s.4d.). Moreover, the Magyar and Szekler population of Transylvania are in possession of ancient electoral rights, and which give to them, at least for some years to come, what almost amounts to universal suffrage. All "Nobles", and they are exclusively Hungarians, and very numerous, who, as such, possessed the franchise in 1872, are permitted to retain it during their life-time. That this privilege affords a great advantage to the Magyars is evidenced by the fact, that 67.4 per cent. of all the electors in Transylvania belong. In the class of "Nobles". Furthermore, the electoral districts have been, since 1868, continually reorganized and readjusted, in order to facilitate, as far as possible, the return of Magyar representatives. In localities inhabited by Magyars, the electoral districts are small and numerous, with perhaps only a few hundred voters in each, while in those parts where the Roumans are in preponderating numbers the districts are extensive, with some thousands of voters. The polling-places are situated, when practicable, in Magyar centres, and I have been assured that many a Rouman voter is more than a day's journey distant from the polling-booths, so that it would be physically impossible for him to exercise his political rights. The Roumans are exceedingly emphatic in their condemnation of the Electoral Law, which they consider to be one of the gravest injustices inflicted on them. They appeal to paragraph 1 of the Law of 1868, incorporating Transylvania into Hungary, in which it is expressly laid down that "all the inhabitants of Transylvania, without distinction of religion, nationality, or language, are hereby declared in enjoyment of equal rights. And the equality of all citizens in Hungary and Transylvania, in civil and political affairs, is hereby guaranteed". They, therefore, demand that the Electoral Laws of Transylvania should be assimilated to those of Hungary, and that a revision should be made of the electoral districts. It certainly seems that the Roumans have on this point a very strong case, and the grievance is one which the intelligent classes, at least, feel very deeply. It may be doubted whether the form they have adopted of giving expression to their dissatisfaction is the wisest, as by abstention from the elections they have left the field open to their opponents, and are likely to achieve little by their silent protest. Many with whom I talked on the subject admitted that the "passivist" policy as it is called, has been a mistake, but they consider that it is too late to reverse it. At the same time, they informed me that little or no benefit was derived from sending Deputies to the Pesth Parliament, where no heed was paid to them, and where indeed, they were subject to persecutions and prosecutions, as in the case of General Doda*.

Complaints on the provincial
administration and judicial procedure

On all questions relating to the provincial administration and to judicial procedure there were many complaints. The Roumans assert that, although they do not follow in local Government the passive policy they have adopted in Parliamentary elections, yet they are unable to procure the election of any reasonable number of Roumans to provincial posts. At first sight I was puzzled with the fact, that in districts in which the Roumans formed 96 or 97 per cent. of the population, there were relatively so few in possession of those posts which are obtained by election. I was told that this was owing to the mode in which the elections were conducted, to the arbitrary distribution and division of the districts, and to the franchise. In the towns the Roumans are usually in the minority. I was assured that throughout the districts of Transylvania and Hungary in which over 2,500,000 Roumans reside, there were only two or three employés of any importance who had succeeded in securing election. There are Roumans on Municipal Councils and other bodies, but in active administrative employement they are comparatively rare. Complaints, I was told, had frequently been made to Parliament, and Petitions presented in respect to abuses at elections, but in no instance had any attention been paid to their remonstrances. As the whole system of elective offices will shortly be abolished by the Bill now before the Hungarian Parliament, further remarks in regard to it may be considered superfluous. The Roumans further declare that they are practically excluded from all posts under the nomination of the Government; and I was shown elaborate statistics proving that the Roumans furnished under 2 per cent. of all the State employés, though numbering about a sixth of the population,

---

* General Doda was a pensioned Austrian General and a Rouman Deputy. He was prosecuted for an electoral address, and was condemned by a lower Court to a fine of 100 l. and to two years' imprisonment. On his petitioning the Throne, further proceedings against him were quashed by order of the Emperor. [A. Nicolson].

and 75 per cent. of this 2 per cent. were in very subordinate offices. I do not think that the small participation which the Roumans are allowed to take in executive and administrative work would be seriously complained of if the laws were administered impartially and equally, and if the rights of the nationalities as secured to them by law were observed.

The provisions of the Law of 1868 are practically a dead letter

The Nationality Law of 1868 provides that all officials should, in their communications with parishes, institutions, private individuals, &c., employ, so far as lies in their power, the language of the interested parties; and that individuals may, in judicial matters, when they have no advocate, employ either their mother tongue or the language generally in use in the locality. The Law further lays down that the Communal authorities must, in their communications with the inhabitants of the Commune, use the language of the latter; and that the State will take care that, in the appointment of judicial and administrative employés, only competent persons shall be selected for service among the different nationalities, and who are perfectly versed in the language in use. There are many other provisions of this Law, passed under the influence and guidance of Déak, for safeguarding the privileges and languages of the nationalities. On all sides, from Saxons as well as Roumans, there was an unanimity of assertion that, with a few isolated exceptions, all the provisions of the Law to which I have referred are practically a dead letter. This would not be denied by the Hungarians themselves; and, indeed, it has frequently been stated, both in Parliament and in the principal Pesth newspapers, that the Law of 1868 was an error, committed in a moment of generous expansion, and that it would be folly on the part of the Government to execute it. Numerous instances were cited to me of Judges being appointed to districts absolutely ignorant of the language of the vast majority of the inhabitants, whereby confusion and needless expenditure were created in all suits brought before the Courts. Interpreters, and generally incompetent ones, had to be hired by the interested parties, as the Judges usually insisted on all the summonses, writs, and other judicial documents being drawn up in Hungarian. A case in which an English Company were engaged in Transylvania was brought to my notice some time ago in Pesth, which illustrates the disadvantages of the present system. The case was tried before a Judge, ignorant both of German and Roumanian, and as all the witnesses on the side of the Company were either Saxons or Roumans, and ignorant of Hungarian, the verdict was given without much attention having been paid to the evidence tendered. I obtained a promise from the Minister of Justice that a new trial should be ordered before a better-qualified Judge, but in most cases this concession is naturally not accorded. It is not only in judicial matters that hardships are experienced, but in all affairs in which transactions have to be conducted with State employés. Irksome and irritating as is the present system, the Roumans look forward with dismay to the time when the Bill for still further centralizing the administration has passed into law, and when a swarm of petty officials will be scattered over the country. I may add that both from Saxons and Roumans there were many complaints as to the want of integrity on the part of the employés, an evil which is, however, so general throughout Hungary, that it can form no special grievance for any nationality.

The measures the Romanians have taken to meet the difficulties

I have now alluded to the chief complaints of the Roumans, and I would beg leave to refer to the measures they have taken to meet the difficulties which they consider to be ever increasing round them. In 1881 the Roumans decided to call together a Conference of Delegates from all districts in Hungary inhabited by Roumans; and this Conference met together at Hermannstadt, and was attended by over a hundred Deputies. In May 1881 the Conference, after several days' sitting, unanimously voted what is termed the "Programme of the Roumanian National party in Hungary and Transylvania", and in which were laid down the objects for which the National party bound themselves to work in a Constitutional and legal manner. The first point in the programme is the recovery of the autonomy of Transylvania; but the leaders of the National party informed me that the insertion of this point in the programme is

Appendices

intended rather as a protest than formulated as a demand, as they are well aware of the impossibility of ever inducing the Hungarians to reverse the Act of Union. It is more a reservation of rights, so that silence on the point may not be interpreted as acquiescence. The programme further demands the legal recognition of the use of the Roumanian language, both in judicial and administrative affairs, in the districts inhabited by Roumans, and that in such districts only those employés should be appointed who were either Roumans by birth, or who had a competent knowledge of the Roumanian language. Further, that the Electoral Law should be remodelled; that the autonomy of the Churches and confessional schools should be secured; and that the Nationality Law of 1868 should be revised and loyally observed. The programme further states that the National party will use their utmost endeavours to resist the tendency to "Magyarize" the nationalities, on the ground that such a tendency brings discord and mutual hatred among the component parts of the State.

The Conference also met in 1884, 1887, and again in 1890, when over 130 Delegates were present, and from a list of the names and occupations of the Delegates which was given to me, it would appear that there were among the Deputies 29 priests, 40 lawyers, 7 bankers and commercial men, 4 journalists, 8 professors, 5 doctors, and the remainder were classified as land-owners and private individuals. In the Conference of 1890, it was decided to adhere to the 1881 programme; and an explanatory declaration of the aims of the National party was made to the Assembly by the President. In this declaration, it was stated that the National party "rejected all tendencies which were working towards an union with Roumania", but that they wished to see a close alliance and a commercial union between the Roumanian Kingdom and the Hapsburg Monarchy.

The declaration also enunciated the suggestion of a Military Convention with Roumania. The National party, the declaration continued, belonged to the Austro-Hungarian Monarchy, and were faithful subjects of the Hapsburg Crown. They desired a prosperous Fatherland, and a strong Austria-Hungary, of which they desired to be good and useful citizens. At the same time, they demanded that at least the laws of the Fatherland should be observed in regard to them. They appealed to all other nationalities to make common cause with them, and the declaration concluded with the observation that the Roumanian element was of importance to the Triple Alliance, and that the more freely the Roumanians were allowed to develop, the stronger would be the dam which they would be able to erect against foreign invaders.

A brief recapitulation of personal impressions: the views of the Romanian peasants

I would beg leave to be allowed to recapitulate as briefly as possible the impression which I received during my recent tour.

While making every allowance for exaggerated statements and a considerable amount of wild and loose talk, I am convinced there is a deep feeling of animosity throughout the educated classes of Roumans against their Hungarian Rulers. As to the views of the peasants I was unable to gather any direct information, owing to the shortness of my stay and my ignorance of their language; but it may be safely assumed, as I was told by many sober-minded and impartial Saxons, that the peasant will follow blindly the orders of his priest, or of his political leaders. At present he is passive and silent; and in many districts where the rich landed proprietors have their estates, and these latter are all on the side of the Government, being in reality Hungarians, it would be imprudent for the peasant to take any very active part in any agitation, Constitutional or otherwise. He, nevertheless, shows, when an occasion arises, and when landlord or official pressure is not exercised, that his sympathies are with the representatives of the National party; and the political leaders count confidently, and, as I was informed from non-Rouman sources, with fair reason, on the support of the peasant whenever it may be necessary to appeal to it.

The Romanians vis-à-vis the Austro-Hungarian Monarchy

I believe that the Roumans are sincere in their assurances, that they desire to remain faithful subjects of the Austro-Hungarian Monarchy, and that they are not looking to a political union with Roumania. Nevertheless, a great strain is being placed on their loyalty to the Hungarian Government; and I much fear that, unless there is some relaxation, of which there seems to be little prospect, that the Roumans may adopt the attitude assumed by the Hungarians before 1867 towards the Austrian Government, and consider Hungary's difficulties as their opportunity. Moreover, they have of late excited a considerable interest in their lot among their brethren in Roumania, and the close neighbourhood and increasing intercommunication cannot fail to have an influence on a people who are daily becoming more dissatisfied with their present position in the Hungarian State.

The various nationalities under Hungarian rule

They cannot refrain from contrasting the situation of the several provinces under the Austrian dominion with that of the various nationalities under Hungarian rule, and they are incensed at the continued efforts of the Hungarians to crush out their national life, and to fuse them forcibly in the crucible of Magyar supremacy.

I had long conversations with moderate, sensible Saxons, and especially with the Lutheran Bishop, Dr. Teutsch, who is a member of the House of Magnates, and well acquainted with the leading politicians and statesmen in Pesth, and he, as well as my other informants, recognized the futility of the present policy of the Hungarian Government. In isolated instances, and among some of the weaker-kneed races, this policy may have occasional successes, but they are convinced that efforts to turn a Rouman into a Magyar must terminate in failure.

Magyar opinions

The Hungarians, on the other hand, assert that the Roumans will never in their hearts be loyal and faithful friends; that they are striving after aims far beyond the modest demands put forth in their published programme; that were the Government to evince a conciliatory and yielding disposition, it would be considered as a sign of weakness; and where an inch was granted an ell would be required. The example of Austria, the Hungarians consider, is but an illustration of the evils of decentralization, and that the want of a strong and firm Central Government has, in Austria, produced not contentment, and peace, and strength, but disorder, and confusion, and weakness. The Hungarians admit, at least the most candid of them, that their policy may be harsh and arbitrary, but if they are to establish a solid Government, and not be swamped by the nationalities around them, the reins must be held in a firm grip; one strong Central Administration, one State language, and, possibly in time, one vernacular, must be imposed, and no segregating influences, such as local autonomy and a Babel of tongues, can be allowed in a State in the situation of Hungary, surrounded by open enemies and insidious intriguers, and requiring all her energies and all her strength to advance onwards.

Conclusions

Time alone can show whether this view is correct. It would be presumptuous in a stranger knowing but the surface of affairs to pass an opinion. At present I can but record the fact, that a large community of over 2,500,000 Roumans are in a condition of serious discontent, and inspired with bitter feelings against their actual rulers, from whom they despair of obtaining any concessions unless extorted by the necessities of the moment; and that this community occupies, in compact masses, a portion of the Hungarian State, which, in possible eventualities, might prove a vulnerable point for hostile attack.

Buda-Pesth, May 25, 1891

Public Record Office, London, F.O. 800/336 (A. Nicolson Papers). South Eastern Europe Confidential Print 2940/1891

APPENDIX E

1895 August 10

The Program adopted by the Congress of the Nationalities held in Budapest

I. The Romanian, Serbian and Slovak alliance declares that it will support the integrity of the lands of the Crown of Saint Stephen in every respect.

II. While respecting the ethnic relations and historic development of Hungary by virtue of which Hungary is not a state whose national character could be determined by any one people alone but only by the totality of Hungarian peoples: Romanians, Slovaks, Serbians, inhabitants of Hungary and Transylvania, while supporting the integrity of the lands of the Crown of Saint Stephen, wish to draw, from the existing ethnic relations and from the historic development of Hungary, all the consequences that are necessary to the administration of the state in respect of the preservation and development of the individual peoples of Hungary.

III. The nature of the Hungarian state, as a result of ethnic relations and historic development, does not allow that one single people, who does not even represent the majority of the population, should assume the right to form by itself the state. Only the totality of the peoples of Hungary have the right to identify themselves with the state, as stipulated in existing laws. Therefore, the so-called idea of a state for one people only, who represents a minority as against the other peoples of Hungary who represent the majority, is in contradiction with ethnic relations and the historic development of Hungary and, at the same time, threatens the living conditions of the other peoples of Hungary, who represent the majority.

IV. In view of this threat which by deeds readies the destruction of the living conditions of the other peoples of Hungary and Transylvania who constitute the majority of the country, the Romanians, Slovaks, and Serbians of Hungary and Transylvania, upholding their own national programs to date, conclude an alliance to protect their nationality in all legal ways and hope that because of common interest the Russians (Ruthenians) and the Germans of Hungary will join this alliance since the alliance will deal with nothing illegal, but is the more so determined to use all legal means in order to place Hungary on its natural foundations. As a natural foundation must be regarded national autonomy within the rounding off of the comitats (counties).

V. This alliance of nationalities does not exclude the possibility that each nationality, on the basis of specific circumstances of its own, may aspire to a development of its own and, for that purpose the allied nationalities promise mutual assistance.

VI. The non-Magyar peoples of Hungary, according to linguistic boundaries, shall be given complete freedom so that the respective autonomous territories (comitats, municipal towns, communes) be imparted the character of the respective nationality through utilization of the national language by administrative and judicial authorities. Wherever there be more than one language, the comitats should be rounded off on the basis of language in the interest of administrative simplification.

VII. The Romanians, Slovaks and Serbians cannot be satisfied in this respect with the so-called law of nationalities of 1868. That law, so it would seem, was enacted only to delude foreigners as to the alleged tolerance toward the nationalities. In reality that law is not observed and is quite illusory. As long as the law of nationalities is in effect, the Serbians, Romanians, and Slovaks will observe it as a law but demand from the rulers of the state that it be respected and, in fact, even broadened in favor of the nationalities and of equal justice. Should the nationalities ever be given the opportunity to be represented in the Hungarian parliament, they shall strive to have the law of nationalities altered in the sense of including the principles of national autonomy contained in this program and, specifically, they shall insist that the obligation of the state to support from its own resources the cultural aims of the various peoples of Hungary be observed in every way in an impartial manner.

VIII. The nationalities of Hungary find themselves in the sad position of not being able to share in the legislature of the Hungarian state. Inasmuch as Hungary is regarded as the domain of one people only, the state power is trying to give the Hungarian parliament the character of its being representative of a national state. Starting from that point of view an unjust and horrendous electoral law was created for Transylvania. In fact, on the basis of the electoral law of Hungary itself, the electorate is formed in such a way and the electoral lists are as a rule so put together that voters belonging to the nationalities appear to be cast aside and crushed in elections. Moreover, the administrative organs use such horrendous pressures and enact such violent measures against the nationalities that these actions are unparalleled in elections held anywhere in civilized Europe. Under these circumstances, when the sacred right of electoral freedom has become illusory, the Serbians, Slovaks and Romanians, as parties, are forced to abstain for the time being from parliamentary elections.

IX. In this connection, the alliance of nationalities seeks the introduction of universal, direct, and secret suffrage, a juster distribution of seats, the removal of administrative pressure at elections and the abolition of those legislative measures whereby failure to pay the tax results in loss of electoral rights.

X. Inasmuch as in Hungary there is no law granting the right of free assembly and association and governmental decrees on these matters are out-of-date and equivocal, administrative organs apply them and explain them in an arbitrary manner so that all possible impediments are used to prevent public meetings and the formation of associations of nationalities, and for these the right of assembly and association is almost non-existent. Thus, the alliance of nationalities demands the enactment of a precise and democratic law regarding the right of free assembly and association.

XI. Inasmuch as under the present system the church and school autonomy granted by law have become illusory through arbitrary governmental interference, the alliance of nationalities demands the respect of the legal church and school autonomy and its enlargement in the true meaning of autonomy.

XII. Considering that in Hungary the great institution of the Jury Court exists exceptionally only for press matters, and taking into account the fact that the composition of such juries is intentionally constituted such that, in matters of the press involving non-Magyars, only Magyars with hostile feelings toward nationalities, and who do not even understand the language of non-Magyars, are passing judgement, which is also proved by the transfer of the Jury Court from Sibiu to Cluj, and since in this way this institution is belied and used against any expansion of freedom of the press of nationalities — which are thus completely deprived of the freedom of the press — the alliance of nationalities requests either the placing of such juries in localities where by the elimination of the detrimental institution of the interpreter, it will be possible to conduct the proceedings in the original language of the press or the total elimination of the institution of the jury and subjection of press actions to ordinary courts.

XIII. Inasmuch as in Transylvania there is an exceptional law of the press and in Hungary the freedom of the press is restricted by the imposition of bail, the alliance of nationalities seeks a unitary law for the unlimited freedom of the press.

XIV. The alliance of nationalities desires freedom in all fields and thus it fights foremost for the freedom of existing religions. From this standpoint the alliance of nationalities will fight by all legal means for the revising of politico-ecclesiastic laws all the more so as those laws are directed against the national life of the various peoples.

XV. The Hungarian nationalities desire that their interests be represented by a minister designated by the Crown in the manner in which Croatia and Slavonia are represented by a minister without portofolio.

XVI. To attain the proposed goals and to achieve a unitary leadership, the alliance of the Slovaks, Serbians, and Romanians will be represented by a committee elected by the representatives of the Serbians, Slovaks, and Romanians, which committee shall include four members from each nation.

XVII. This committee shall avail of all means for its organization and shall see to it that, on all occasions, it will lodge periodical protests against the policy of denationalization carried out by the present holders of state power.

XVIII. The committee shall take care that good understanding among the various nationalities shall be safeguarded and that everything that may detract from such good understanding shall be kept at bay.

XIX. To elucidate the European public at large, which is not sufficiently acquainted with national relations in Hungary, the committee shall strive to inform the European press.

XX. The congress decides to meet periodically and entrusts the central committee with the establishment of the venue and date and with the calling of meetings.

XXI. The convening committee shall elaborate even before the next meeting of the congress a Memorandum which, on behalf of all peoples represented by this alliance, shall fully describe the situation with a view to laying it before His Majesty.

XXII. In the event that the meeting of the delegated committee shall be brought to naught, the principles incorporated in this program, shall be the directives for the national policy of the Slovaks, Serbians, and Romanians".

Telegraful Român (Sibiu), XLIII, 1895, 15 August, No. 86. [In Romanian]

APPENDIX F

1 December 1918

Decision on the Union of Transylvania with Romania

I. The National Assembly of all the Romanians of Transylvania, the Banat and Hungary, meeting through their elected representatives at Alba Iulia on November 18/December 1, 1918, proclaims the unification with Romania of these Romanians and the territories inhabited by them.

II. The National Assembly secures the afore-mentioned territories provisional autonomy until the meeting of the Constituent Assembly elected by universal suffrage.

III. In this connection, the National Assembly proclaims the following as fundamental principles governing the new Romanian state:

    1. Full national freedom to all the coinhabiting peoples. Each people will have the right to education, administration and justice in its own language through individuals from among them; each people will be entitled to representation in the legislative bodies and to participation in the government of the country according to its numerical size.

    2. Equal rights and full religious freedom for all denominations in the state.

    3. Full implementation of a truly democratic system in all the fields of public life: universal, direct, equal and secret suffrage (proportionally by communes) for all male and female voters over 21 years of age and representation in communes, counties and Parliament.

Appendices 167

    4. Full freedom of the press, assembly and association; freedom of propagating all human ideas.

    5. A radical land reform, with all the estates, especially the large ones, being registered. On the basis of this registration and the abolition of fidei-comissions on the strength of the right to divide, as requested, the large estates, the peasant will have his own property (arable land, grazing land, forest) within at least the limits of his and his family's labouring possibilities The ruling principle of this agrarian policy is to encourage, on the one hand, a social levelling off and on the other, a rise in production.

    6. Industrial workers are granted the same rights and advantages which are enforced in the most advanced industrial states in the West.

IV. The National Assembly expresses its desire that the Peace Congress should lay the foundations of a community of free nations where justice and freedom should be enjoyed by all nations, large and small alike, that war as a means of settling international relations should be avoided.

V. The Romanians gathered at this National Assembly welcome the liberation of their Bukovinian brethren from under the yoke of the Austro-Hungarian monarchy and their unification with Romania, their motherland.

VI. The National Assembly welcomes with affection and enthusiasm the liberation of the Czechoslovak, Austro-Hungarian, Yugoslav, Polish, and Ruthenian nations, formerly ruled by the Austro-Hungarian monarchy, and wants this to be made known to all these nations.

VII. The National Assembly piously honours the memory of those courageous Romanians who shed their blood in the war in order to realize our dream and who gave their lives for the freedom and the unity of the Romanian nation.

VIII. The National Assembly expresses its satisfaction with and admiration for the allied powers which saved civilization from the claws of barbarity by their resounding and tenacious battles against an enemy prepared for war for many decades.

IX. To carry on with the affairs of the Romanian nation of Transylvania, the Banat and Hungary, the National Assembly decides the election of a Romanian Grand National Council, entitled to represent the Romanian nation anytime, anywhere, before all the nations of the world and to take any decision deemed necessary for the good of the nation.

Romanian text in *Românul*, Arad, VII, no 20 of 20 November/3 December and *Telegraful român*, Sibiu, LXVII, no 131 of 23 November/6 December 1918. English translation by Sanda Mihailescu in M. Mușat and I. Ardeleanu, *Political life in Romania 1918—1921*, București, 1982, pp. 20—21.

INDEX

Adamclisi, Tropaeum Trajani at, 2
Ady Endre, 64
Ahtum, Dux, 7, 10
Alba, 3, 39, 122
Alba Iulia, 4, 19, 23, 26; National Assembly of 1918 at, 66, 166—168
Angevin kings, 28
Antalffy, Endre, 65
Appony, Count Albert, 57, 58
Approbatae Constitutiones, 21, 122—124, 126—129
Apulum, 4
Arad, county 34; town, 64, 145
Armenians, in Transylvania, 125, 128
Aromanians, in Transylvania, 30
Arpad, 6, 15
ASTRA (Transylvanian Literary Association), 149
Atanasie, Anghel, 26
Aurelian, Emperor, 2, 3
Ausgleich (Compromise of 1867), 52, 53
Austro-Hungarian Monarchy, 31, 32, 63, 64, 68, 70, 157—159, 167
Avars, 4, 10
Axis powers, 77, 79

Bălcescu, Nicolae, 44, 45, 50
Balkan Peninsula, 20, 24, 83
Banat, vi, 3, 6, 16, 19, 25, 43, 44; in 1918, 64, 66, 68; as Valachia Cisalpina, 29

Bánffy, Dezsö, 57
Barițiu, George, 46, 147
Bărnuțiu, Simeon, 46, 48, 49
Bartók, Béla, 64
Báthory, Andrew, 11
Báthory, Sigismundus, 120
Beiuș, 148
Béla, king, 6, 118, 119, 125
Bethlen, Count István, 74
Bihor, 6, 10, 34
Bîrsa, 11, 30
Bistrița, region, 11; town, 12, 78
Black Sea, vi, 1, 6, 11, 17
Blaj, 54, 145; Assembly of, 48, 49, 60
Blasendorf, *see* Blaj
Bobîlna, uprising of, 18
Bogdan, voevode, 16
Bonfini, Antonio, 23, 121
Bonomi, A., 84
Brașov, town, 35, 79, 94; district, 135
Buccow, Adolf von, 30
Bucharest, 14, 67, 82, 144
Buda, 11, 29; Pashalik of, 19
Budapest, 55, 61, 63, 64, 80; *see also* Buda and/or Pest
Bukovina, 52, 167
Bulgarians, in Transylvania, 125
Bulgars, 6, 7
Burebista, 1
Buteanu, Ioan, 50
Byzantine Empire, 6, 19; Byzantine Church, 10; *see also* Eastern Roman Empire

169

Caesar, Caius Julius, 1
Caioni, Ioannes, 17
Calvinist Church, see Protestant Church
Capodistrias, Count, 24
Căpîlna, Union of, 18
Caransebeş, 145
Carei, 82
Carol II, 77
Carpathians, vi, 5, 6, 7, 10, 12, 35, 43, 45, 48, 50, 52, 54, 59, 82
Catherine II, of Russia, 24
Catholic (Roman-Catholic) Church, 10, 20, 121, 129, 130; proselytism, 26, 30, 35; see also Western Church
Ceauşescu, Nicolae, 98
Cetatea de Baltă, 12
Christianity, 10; in Dacia, 3; in Transylvania, 3
Churchill, Winston Spencer, 79
Ciano, Count Galeazzo, 77
Ciceu, 12
Cloşca, 39
Cluj, 3, 18, 39; Diet of, 25, 142; Memorandum Trial at, 61; University, 149; Jury Court, 150, 164; in 1940, 78, 79; see also Cluj-Napoca
Cluj-Napoca, Magyar higher education etablishments at, 91—94, Magyar Opera House, 96
Cluj-Mănăştur (Colos-Monostra), 119, 124, 134
Constantinople, 17, 119
Coresi, 35
Corvinus, 17; John (Ioannes) Corvinus of Hunedoara, 17, 18, 120, 121; Matthias Corvinus, 18, 23, 120, 121
Covasna, county, 89, 90
Crăciun, Gheorghe, 34

Crişan, 39
Crişana, vi, 19, 31, 52; in 1918, 64, 66
Croatia, 9, 19, 25, 58, 165; Croat nationality 61, 144
Csuták, Kálmán, 50
Cumans, 4, 7
Curzon of Kedleston, Lord, 68
Cuza, Ioan Alexandru, 60
Czechoslovakia, Czechoslovak gvt., 67, 68, 74, 75, 76, 167

Dacia, Dacians, vi, 1—4, 12, 117, 118; ideal of revival, 22, 23, 24, 41, 43
Daco-Romans, 1—4
Danube, ii, 2, 9, 17, 20, 22, 30, 39, 118
Deák, Ferenc, 56, 155
Decebalus, 1
Dej, 18, 78
Diaconovich, Corneliu, 151
Diets:
    Hungarian: at Pest, 34, 53; at Pressburg 48
    Transylvanian: 13, 20, 36, 43, 52, 54, 122, 123; of 1744, 130, 132; of 1841—1842, 45, 138; of 1848, 48; of 1865, 53; of Sibiu, 53, 54, 55
Dinnyés, Lajos, 85
Dobrogea (Dobrudja), vi
Doda, General Traian, 154
Dozsa, George, 34

Eastern Church, 117, 118, 119
Eastern Roman Empire, 3, 118; see also Byzantine Empire
Education:
    in Transylvania: before 1868, 147; after 1868, 58, 147, 148, 150; see also Magyarization
    in Romania, 72, 73, 91—98
Eötvös, Baron József, 56

Index

Europe, 42, 79; European press, 40, 41; Romanians' memoranda addressed to, 61, 144, 151, 165; South-East, 24; Central, 25, 74, 83

Făgăraș, 7, 12, 39, 128
Fay, Dr. J., 67
Ferdinand I, of Habsburg, 19, 29, 121, 141
Florence, Council of, 26
France, French, 40, 41, 47, 52, 68, 74, 84, 104
Fraterna Unio, see Unio trium Nationum

Galicia, 47, 140
Gasperi, Alcide de, 84
Gelu, Dux Blachorum, 6, 7, 118, 125
Gepids, 4
Germans:
  in Transylvania, before 1918, 63, 125—126, 162; see also Saxons, Swabians
  in Romania, 86, 87, 94, 95, 96, 97
Germany, 74, 75, 77, 83, 149
Getae-Dacians, 1
Glad, Dux, 6, 7, 10
Gömbös, Gyula, 75
Golescu-Negru, Alexandru, 50
Great Britain (England, Angleterre), 40, 68, 74, 83, 84, 100, 104, 110, 156
Gritti, Aloisio, 24
Groza, Dr. Petru, 84, 85
Gyla, Duke, 10
Gyula, the Elder, 119; the Younger, 119

Habsburg Monarchy (Austria, Court of Vienna, Austrian Empire): rule in Transylvania, 13, 24, 25, 52, 54, 125, 126, 141, 143; passim
Haemus (Balkans), 1
Harghita, county, 89—90
Hațeg, 7, 12
Hermannstadt, see Sibiu
Hitler, Adolf, 75, 77—79, 82, 83
Holy League, 25
Horea, 39—41, 99, 101—104, 110, 111, 114
Horthy, Miklós, 74—77, 80
Hull, Cordell, 79
Hunedoara, 35, 36, 120
Hungarians in Romania, see Magyars
Hungary, 9, 12, 19, 25, 32, 52, 54; after 1868, 55—62, 142—146, 152—159; in 1918, 63—65; after 1918, 64, 69, 74, 75, 76, 77, 80, 85
Huns, 4
Hunyadi, see Corvinus

Iancu, Avram, 49, 50
Illyrian nation, 136, 137; see also Serbs
Iorga, Nicolae, 78
Italy, 40, 52, 62, 68, 74, 77; Italians under Habsburg rule, 54

Jakó, Zsigmond, 98
Jászi, Oszkár, 65
Jews, 3; in northern Transylvania 81
Joseph II, Emperor, 31, 38—40, 42, 99—116, 129—130

Karlowitz, Peace of, 13
Kodály, Zoltán, 64
Kossuth, Louis, 46, 60
'Kossuth' Radio Station (USSR), 83
Kun, Béla, 67

Lachmann, F., 62
League for the Cultural Unity of All Romanians, 144
League of Nations, 71, 74, 76
Leopold I, Emperor, 26, 36, 125, 126, 129; Leopoldine Diplomas, 25, 26, 36
Leopold II, Emperor, 117, 133
Lépes, Lorand, 120
Locarno, Treaty of, 75
London (Londres), 79, 110
Louis the Great, of Hungary, 16
Lukaris, Patriarch Cyril, 34, 35
Lugoj (Lugos), 145
Lutheran Church, see Protestant Church

Magyarization, 33, 45 ff., 55, 58 ff., 138—139, 143 ff., 149 ff., 157, 159 ff.
Magyars, Hungarians:
in Transylvania before 1918: passim; arrival in, 6, 8, 118; Anonymus on, 7, 119; religion, 10, 20, 63, 121—122, see also Catholic, Protestant Church; census, 30, 31, 32, 47, 140; as "political nation", see Unio Trium Nationum, Nobles, Diets in Romania, 86—87 ff.; socio-economic developments 88 ff.; culture 89 ff.; education, 91—93; science, arts and literature, 93 ff.; mass-media, 94 ff.; religion, churches, 97—98
Maior, Petru, 37
Maramureş, vi, 7, 16, 19, 34, 39, 52; in 1918, 64, 66
Maria Theresa, 30, 129, 142, 145
Méliusz, József, 94
Menumorut, Dux, 6, 7
Michael, the Brave, 11, 21, 22, 23
Micu Clain, Bishop Inochentie, 27, 36, 37, 42

Micu Clain, Samuil, 37
Miercurea Ciuc, 78, 89
Millerand, Alexandre, 69
Mohács, battle of, 19, 141
Mohammed, the Conqueror, 17
Moldavia (Moldova), vi, 4, 10, 11, 12, 13, 16, 17, 60; passim
Munich, Agreement of, 75
Munkács, 30
Mureş, vi, 8; county, 1, 3, 4, 89 ff.
Murgu, Eftimie, 44
Mussolini, Benito, 74, 75

Napoca (Cluj-Napoca), 4
National Assembly: Romanian desideratum in 1791, 43, 135, 136
Nation, Nationality:
'political nation', 20, 21; see also Nobles, Unio trium Nationum idea of, among Romanians, 21, 35, 36, 37, 41, 42, 44, 45 ff., 60, 61, 144, 157, 158, 160; Supplex Libellus, 117—137
Nationalities in Monarchy:
statistics and distribution, in Transylvania, 28 ff., 135; in Hungary and Transylvania, 32, 33, 63, 140; see also Magyars, Szeklers, Saxons
non-Magyar, 32, 47, 56, 57;
Law of Nationalities, 56, 58, 143, 148, 155 ff., 162; Congress of (1895), 161—165
Nationalities coinhabiting in Romania:
number v, 86; status of, 86—90; cultural policy, 91—97
Nicolson, Arthur, 61, 62; his Report, 140—160
Nobles, Nobility, 15: 'political nation', 20, 21; Magyar nobility/ nation, 21, 23, 45; in the 'Fraterna Unio', 121, 122, 127; see also Diets

Index

Odorheiul Secuiesc, 89
Olahus, Nicolaus, 17, 23, 120
Olsavsky, Bishop Manuel, 30
Oltenia (Little Wallachia), vi
Oradea, 78, 81, 145, 148
Orăștie, 1, 120
Orthodox (Greek-Oriental) Church, in Transylvania, 15, 16, 20, 26, 28, 34, 35, 104, 105, 122—123, 126—130, 134, 142, 145, 146; census, 30, 31
Ottoman Empire (The Porte, Turks), 10, 13, 17, 22, 24, 25; suzerainty over Transylvania, 12, 19, 20; over Hungary, 19, 141

Pannonia, 6, 9, 28
Papacy, 22, 25
Paris, Peace Conference (1919), 68, 70; Peace Treaty of 1946, 84
Partium, 19, 25
Peasantry, see especially 18, 34, 38—41, 51, 99—116, 132, 133, 134; Nicolson on ~ 158
Pechenegs, 4, 6
Pennsylvania, 110, 115
Pest (Pesth), 34, 38, 59, 143, 149, 152, 153, 155, 156
Petru Rareș, voevode, 10, 12
Poland, Polish, 25, 47; Polish nation, 167
Potaissa (Turda), 4
Pozsony, 19
Pragmatic Sanction, 142
Preiss, Baron von, 31, 39
Press, journalists, 40, 61, 100—105, 107; condition of, in Transylvania, 55, 150, 151, 152; in Romania, 72, 73; 93—97
Pressburg, 19
Programs: see especially, in 1784, 109, 111; in 1791, 134; as formulated by Bishop Inochentie, 36—37; in 1847, 44; in 1868, 54; in 1881, 156—157; in 1890, 157—158; in 1895, 161—165; in 1918, 166—168
Protestant Church, in Transylvania, 128; Calvinist (Reformed), 20, 26, 34, 121, 126; Lutheran (Evangelical), 20, 34, 121, 126; Unitarian (Socinian), 20, 30, 121, 126; census, 30, 31

Rákóczi, Francis II, 34
Rákóczy, George, 122
Revisionism, Magyar, 74—77
Ribbentrop, Joachim von, 77
Roman civilization, in Dacia, 2, 3, 4
Romania
   before 1918, and Transylvania, 61, 62, 144, 149, 157, 158; Union of Transylvania with, 63—70, 166—168;
   in interwar period, 72, 74, 75, 76, 78, see also Vienna Diktat of 1940;
   liberation of northwestern Transylvania, 82—85; Socialist Romania: and the status of coinhabiting nationalities, 87—90; cultural policy, 91—97
Romänische Revue, 147, 150, 151
Romanul, Metropolitan Miron, 146
Roth, St. L., 46, 138—139
Russia (Tsarist Empire), 24, 25
Ruthenians, 7, 30, 162; Ruthenian nation (1918), 167

Sarmizegetusa, 1, 2
Saxons:
   in Transylvania: settlement, 8, 10, 11, 15, 125; numbers 29, 30, 31, 32, 140; religion, 30—31, see also Protestant Church; as 'political nation', see Unio trium

nationum; Romanians and, 132, 134, 138—139, 150
   in Romania after 1918; 62, 67; *see also* Nationalities coinhabiting in Romania
Satu Mare, 78, 82
Serbia, 19, 62, 68
Serbs, in Monarchy, Serb nationality, 53, 56, 61, 144, *see also* non-Magyar nationalities
Sibiu, 3, 11, 12, 25, 28, 39, 62, 120, 145, 149, 150; *see also* Diet
Sighet, 78, 81
Simeon Ştefan, Metropolitan, 35
Simonffy, József, 49
Slavs, 4, 7
Slovaks, Slovak nationality, 19, 53, 56, 58, 161—165; Slovakia, 75, 76
Soviet Union, 76, 83, 84
Stephen the Great, 10, 20
Suleiman, the Magnificent, 19
Swabians, in Banat, 68
Szabó, T. Attila, 95
Szeklers:
   in Transylvania: settlement in, 8, 10, 28; numbers, 29, 30, 31, 32, 140; religion 30—31; as 'political nation', *see* Unio trium Nationum; Romanians and, 10, 13—14, 28, 29;
   in Romania, 67, 87, *see* Nationalities coinhabiting in Romania

Şaguna, Metropolitan Andrei, 145
Şincai, Gheorghe, 37

Tatars, 4, 10
Tardieu, André, 68
Teutsch, Dr. Friedrich, 62, 159
Timişoara, 94

Tîrgu Mureş, 65, 78, 84, 91—94, 96
Tisza, Kálmán, government, 57, 58
Titulescu, Nicolae 71, 74, 75
Trajan, Emperor, 2, 23, 117
Transylvanian School (Şcoala Ardeleană), 37, 43
Trefort, School Laws of, 58
Trianon, Peace Treaty of, 69, 70
Triple Alliance, 158
Tuhutum, Dux, 7, 118, 119, 125, 126, 128, 129, 131
Turda (Potaissa), 4
Turkic peoples, 10

Uniate (Greek Roman Catholic) Church, 26, 37, 130, 142, 144, 145; census, 30, 31
Unio trium Nationum (Fraterna Unio of 1437), *see especially* 18, 25, 120, 121, 124, 127, 132; *also* Diets
Unitarian (Socinian) Church, *see* Protestant Church
United States, 40, 100, 110, 115, 116; to the Peace Conference, 68; on Vienna Diktat, 79, 83, 84; on Romania's cultural policy, 97

Vaida-Voevod, Alexandru, 63
'Valachie', as Transylvania, 110
Vandals, 4
Varlaam, Metropolitan, 35
Varna, battle of, 17
Vienna, city: Transylvanian Court Chancellery at, 13, 25, 52, 62; Parliament at, 142; University, 149
Vienna Diktat (Second Arbitration Award, of 1940), 71, 77—81, 83, 84

Index

Visigoths, 4
Voevode, Romanian Voevodeship, 6, 7, 9, 19—20

Wallachia (Țara Românească), vi, 4, 11, 13, 16, 17, 19, 20, 22, passim
Werbőczy, István, 34
Wesselényi, Miklós, 45, 46, 47
Western Church, 119, 121, 134

Wilson, Woodrow, 66; principles of, 65
World War I, 63, 68, 71

Yugoslav nation, 167; Yugoslavia 67, 74, 76

Zapolya, John, 19, 141
Zürich, Peace of, 39

INDEX
OF QUOTED AUTHORS AND WORKS

Anonymus (P. Dictus Magister), 6, 7, 118, 119, 125
Approbatae Constitutiones, 21, 122—124, 126—129
Armbruster, Adolf, 23, 67

Babinger, Franz, 29
Bălcescu, N., 45
Bányai, László, 65, 78, 80, 81
Barițiu, G., 55
Bărnuțiu, S., 49
Benkö, Joseph, 30
Benoist, Charles, 57, 58
Bernath, Mathias, 30, 31, 38, 39
Bielz, E. A., 32
Bodea, Cornelia, 47, 49, 60
Bogdan-Duică, G., 49
Boner, Charles, 60
Bowen, Francis, 50
Brătianu, G. I., 17
Brève histoire de la Transylvanie, 66
Brissot, J.-P., 40, 41, 99, 116

Ciobanu, Virgil, 31
Cholnoky, Jenö, 12
Csatári, Daniel, 78, 80, 81

Deák, Imre, 50
Decei, Aurel, 20
Demény, Lajos, 10
Desbaterile Adunării Deputaților, 67
Din istoria Transilvaniei, 60

Diner-Dénes, J., 58
Documenta Historiam Valachorum 16
Documente privind istoria României, vi
Documents on German Foreign Policy, 77, 78
Dragomir, Silviu, 49, 71
Dücso, Csaba, 80, 81

Eckhart, Ferenc, 9
Edroiu, M., 40
Ekrem, Mehmet Ali, 4
Eliade, Mircea, 1

Fellner, Fritz, 69
Fényes, Elek, 32, 33
Ferenczi, Z., 47
Ficker, A., 32

Gazeta Ardealului, 65
Gazeta poporului, 64
Ghica, Ion, 50
Göllner, Carol, 67, 139
Gromo, Giovanandrea, 29
György, Lajos, 73

Hain, J., 32
Herodotus, 1
Hildebrandt, Conrad Jacob, 29
Histoire de la Hongrie, 57
Hitchins, Keith, 50, 61
Holban, Maria, 16

Index

Homan, Bálint, 12
Horváth, Zoltán, 57
Hunfalvy, P., 32
Hurmuzaki, Eudoxiu de, 16, 22, 29

Iorga, N., 22, 29

Jaeger, H. F., 67
Jászi, Oszkár, 65

Kann, Robert, 53
Kéza, Simon of, 7, 28
Koppándi, Sándor, 10
Kovács, Endre, 44
Kurze Geschichte der Rebellion in Siebenbürgen, 41

Látóhatár, 78
Leeper, A. W. A., 70
Léger, Louis, 53
Lescalopier, P., 23, 24
Lukinich, E., 14
Lupaș, I., 35

Macartney, C. A., 56, 57
A Magyar korona országainak 1910 évi népszámlalása, 63
A Magyar munkásmozgalom története, 67
Magyar statistikai évkönyv, 58
Magyar statistikai közlemények, 59
Magyarország közoktatás ügye, 58
Manuilă, Sabin, 63
Marczius Tizenötödike, 48
Marea Unire de la 1 decembrie 1918, 64
Matei, Ion, 20
May, Arthur J., 54
Mihalyi, I., 16
Moga, I., 11, 12, 13
Monumenta medii aevi historica Poloniae, 17

Nägler, Thomas, 11
Nagy-Talavera, Nicholas M., 81

Nagyajtay Cserei, Mihály, 14
Naumann, Friederick, 33
New York Times, 79
Nicolson, A., 61, 140—160
North American Review, 50

Oguzname, 4
Orientalia Christiana Periodica, 30

Păcățian, T. V., 49
Papacostea, Șerban, 16
Papiu Ilarian, Al., 60
Pascu, Șt., 38
Prodan, D., 18, 36, 40, 47, 137

Real-Zeitung, 40
La Réforme agraire en Roumanie, 72
Reicherstorffer, Georg, 23, 28, 29
Revista Cultului Mozaic, 81
Românul, 66, 168
Roth, Stephan Ludwig, 46, 138, 139
La Roumanie, 70
Rutkovski, Ernst R. V., 62

Sassu C., 31
Scriptores rerum Hungaricarum, 28
Seconde lettre d'un défenseur, 41, 99—116
Seton-Watson, R. W., 33, 55, 57, 70, 74
Siebenbürgisch Deutsche Tageblatt, 62
Söllner, J., 31, 32
Someșan, L., 11
Supplex Libellus Valachorum, 21, 117—137
Szentpétery, I., 28

Szilágy, Sándor, 14
Szöke, Mihály, 14

Telegraful român, 168
Telkes Simon, 59
Titulescu, Nicolae, 75
Török-Magyarkori Allam-Okmány-
    tár, 35
I. Tóth, Zoltán, 57
Tröster, Johann, 23, 29

Verancsics, Antonius, 29
Veress, A., 29
Wesselényi, Miklós, 45—47

Wessely, Kurt, 26

Zimmermann—Werner, 15
Zimmermann—Werner—Müller, 16

LIST OF ILLUSTRATIONS

1. Painted vessel from the Neolithic Age. Cluj-Napoca, Transylvania History Museum.
2. Spiral bracelet, 13th c.B.C. From M. Petrescu-Dîmbovița, *Depozitele de bronzuri din România*, București, 1977, p. 259.
3. Dacia in the time of Burebista. Map by A. Vulpe, in D. M. Pippidi (ed.), *Dicționar de istorie veche a României*, București, 1976, pp. 88—89.
4. Dacian peasant. Rome, Vatican Museum.
5. Dacian nobleman. Naples, National Museum.
6. Fight between Dacians and Romans. Rome, Trajan's Column.
7. Dacians'return after the war. Rome, Trajan's Column.
8. Roman Dacia, Map, in C. C. Giurescu and D. C. Giurescu, *Istoria românilor*, București, 1971, map no. 3.
9. The migration period on Romania's territory. *Ibid.*, map no. 4.
10. Page from Anonymus, *Gesta Hungarorum*. Vienna, Osterreichische National Bibliothek.
11. Modern transcription of the same page. From Ștefan Pascu, *Voievodatul Transilvaniei*, 2nd ed., II, Cluj-Napoca, 1972, p. 25.
12. Christian votive inscription. Bucharest. History Museum of the Socialist Republic of Romania.
13. The battle of Posada (1330). Miniature from *Chronicon pictum vindobonense*. Vienna, Oesterreichische National Bibliothek.
14. The church of Densuș (Hunedoara county), 13th c.
15. The church of Strei (Călan, Hunedoara county), 13th c.
16. The political-administrative organization of Transylvania in the 14th c. From Ștefan Pascu, *op. cit.*, pp. 220—221.
17. Nicolaus Olahus (1493—1568). Wood-engraving by Hans Seebald Lautensach, 1560. From Bucko Voitech, *Mikulas Olah a jedo doba*, I, Bratislava, 1940, p. 3.
18. Johannes Honterus (1498—1549). Anonymous engraving, 17th c., Sibiu, Brukenthal Museum.
19. John Corvinus of Hunedoara (?—1456). Anonymous engraving, 18th c., Bucharest, BARSR (Library of the Academy of the Socialist Republic of Romania).

20. Prince Michael the Brave (1593—1601). Engraving by Aegidius Sadeler. Bucharest, BARSR.
21. Bran Castle, 14th c.
22. The Black Church, Brașov, built 1384—1477.
23. Horea (1784). Oil painting by Johann Martin Stock. From O. Beu, *Răscoala lui Horia în arta epocei*, București, 1935, pl. 12.
24. Cloșca (1784). *Ibid.*, pl. 47.
25. Crișan (1784). *Ibid.*, pl. 47.
26. Bishop Inochentie Micu Clain. Bucharest, BARSR.
27. Title page of the *Supplex Libellus Valachorum*, Claudiopoli, 1791. Bucharest, BARSR.
28. Gheorghe Șincai (1754—1816). Bucharest, BARSR.
29. Petru Maior (1761—1821). Bucharest, BARSR.
30. Simeon Bărnuțiu (1808—1864). Lithograph. E. Sieger, Vienna, Bucharest, BARSR.
31. Nicolae Bălcescu (1819—1852). Oil painting by Gheorghe Tattarescu. Bucharest, Art Museum of the Socialist Republic of Romania.
32. The Great Blaj Assembly of 3/15 May 1848. Watercolor by I. Petcu. Bucharest, BARSR.
33. The champions of the Romanian nationality of Transylvania (1848—1849). Lithograph after Barbu Iscovescu. Paris. Bucharest, History Museum of the Socialist Republic of Romania.
34. Metropolitan Andrei Șaguna (1809—1873). Sibiu, Romanian Orthodox Metropolis.
35. Stephan Ludwig Roth (1796—1849). Daguerreotype, Würtemberg, 1845. Bucharest, BARSR.
36. Ludovic Mocsáry. Editorial in *Familia* (Budapest, 15, 1879). Bucharest, BARSR.
37. The Romanian Memorandists (1892—1894). Bucharest, BARSR.
38. Romanian politicians who conducted negotiations with Count Tisza, 1910. Bucharest, BARSR.
39. The National Assembly of Alba Iulia, 1 December 1918. Bucharest, BARSR.
40. The Coronation Church of Alba Iulia, built 1921—1922. Architect Victor Ștefănescu.
41. Hungary according to the fears of Kossuth in 1850. Map and caption from Oscar Jászi, "Kossuth and the Treaty of Trianon", *Foreign Affairs*, vol. 12, no. 1, October 1933, p. 92.
42. Hungary according to the Treaty of Trianon, 1920. *Ibidem.*
43. Octavian Goga (1881—1938). Bucharest, BARSR.
44. Liviu Rebreanu (1885—1944). *Ibidem.*
45. Victor Babeș (1821—1907). One of the founders of modern microbiology. *Ibidem.*

46. Traian Vuia (1872—1950). Romanian inventor, pioneer of aviation: first flight in a self-powered machine of his own make, 1906. *Ibidem*.
47. The Romanian Lycée in Blaj. Architect Victor Smigelschi.
48. The Polytechnical Institute of Timişoara. Architect Duiliu Marcu.
49. The University House of Cluj-Napoca. Architect George Cristinel.
50. Interior of the National Theater of Timişoara. Architect Duiliu Marcu.
51. The monument of Păuliş (Arad county). Architect Miloş Cristea; sculptors Emil Vitroel and Ionel Muntean.
52. The iron-and-steel mills of Reşiţa (Caraş-Severin county).
53. The National Theater of Tîrgu-Mureş. Architect C. Săvescu